RESOLUTE AND UNDERTAKING CHARACTERS:
THE LIVES OF WILHELM AND OTTO STRUVE

ASTROPHYSICS AND SPACE SCIENCE LIBRARY

A SERIES OF BOOKS ON THE RECENT DEVELOPMENTS
OF SPACE SCIENCE AND OF GENERAL GEOPHYSICS AND ASTROPHYSICS
PUBLISHED IN CONNECTION WITH THE JOURNAL
SPACE SCIENCE REVIEWS

VOLUME 139

RESOLUTE AND UNDERTAKING CHARACTERS:

THE LIVES OF WILHELM AND OTTO STRUVE

by

ALAN H. BATTEN

Dominion Astrophysical Observatory
Herzberg Institute of Astrophysics
Victoria, B.C., Canada

D. REIDEL PUBLISHING COMPANY

A MEMBER OF THE KLUWER   ACADEMIC PUBLISHERS GROUP

DORDRECHT / BOSTON / LANCASTER / TOKYO

Library of Congress Cataloging in Publication Data

Batten, Alan Henry, 1933–
   Resolute and undertaking characters : the lives of Wilhelm and Otto Struve
/ by Alan H. Batten.
   p.   cm. — (Astrophysics and space science library ; v. 139)
   Includes bibliographies and indexes.
   ISBN 978-94-010-7798-9   ISBN 978-94-009-2883-1 (eBook)

   DOI 10.1007/ 978-94-009-2883-1

   1. Struve, F. G. W. (Friedrich Georg Wilhelm), 1793–1864. 2. Struve, O. W.
(Otto W.), 1819–1905. 3. Astronomy—Soviet Union—History—19th century.
4. Astronomy—Soviet Union—History—20th century. 5. Astronomers—Soviet
Union—Biography. 6. Astronomers—Germany—Biography. I. Title. II. Series.
QB36.S75B38   1987
520′.92′2—dc 19
[B]   87–30386
                                                                                 CIP

Published by D. Reidel Publishing Company,
P.O. Box 17, 3300 AA Dordrecht, Holland.

Sold and distributed in the U.S.A. and Canada
by Kluwer Academic Publishers,
101 Philip Drive, Norwell, MA 02061, U.S.A.

In all other countries, sold and distributed
by Kluwer Academic Publishers Group,
P.O. Box 322, 3300 AH Dordrecht, Holland.

To
Lois
who walked with me
in the footsteps of the Struves

TABLE OF CONTENTS

Do you remember the tall Bernhard? He left home in January month to go to IRKUZK IN SIBERIA, in advance of 4,500 miles from us. A friend of mine, General Muraviev, recently became Governor General of the Eastern Siberia, and took him home to his own office under very convenient conditions. You can believe that this was an arrangement not made by me the father, but in consequence to the wishes and the spontaneous declaration of the young man, with which I was very satisfied, for I like a resolute and undertaking character.

From a letter (in English) written by Wilhelm Struve to George Airy (Astronomer Royal) on April 20/8, 1848.

WILHELM STRUVE

OTTO STRUVE

PREFACE

My interest in the history of the Struve family is long-standing but lay dormant until 1972, when I found myself organizing a symposium of the International Astronomical Union in memory of the second Otto Struve. To satisfy my own curiosity, I investigated the precise relationships of the famous astronomers in the family and published an account of them, based mainly on secondary sources. The exercise made me aware that there was no biography in English of the first and probably still the greatest astronomer in the clan -- Friedrich Georg Wilhelm Struve. Wilhelm's son, the first Otto, wrote an account (in German) of his father's life, intended primarily for family and close friends and --though printed-- not generally available. Through the kindness of a family member I have a copy from which I have been able to work. The Soviet historian of science, Z.K. Sokolovskaya, wrote a biography in Russian, in 1964, to mark the centenary of Wilhelm's death. This had a limited edition, and my efforts to obtain a copy failed. Neither work has, in its entirety, been translated into English, although Michael Meo of Oakland, California, and Kevin Krisciunas of Hilo, Hawaii, have kindly made available to me their unpublished translations of some sections of the latter. In the absence of a complete copy, however, when I decided to attempt an English-language biography, I thought it best to do so independently of Sokolovskaya's.

Wilhelm Struve deserves a biography solely on account of his own contributions to astronomy and geodesy, but his fascination is increased by the number of descendants who became eminent in astronomy (and in other scholarly fields) and by the turbulence of the times through which he lived. There must have been something extraordinary about the man and I soon came to feel I could not do him justice in a book that attempted to cover the whole family saga. On the other hand, the career of his son, the first Otto, was too closely bound to his own to be separated. In turn, Otto's life was so long that he saw the most active phases in the lives of *his* two astronomer sons. Thus the book now completed deals with three generations of astronomers in the family. If ever a sequel is written, it will deal primarily with the lives and work of two cousins: Georg and the second Otto.

Although I hope this book has some value as history, my aim is to reach a wider group than historians of astronomy. Too often, scientists are the object of either uncritical adulation or indiscriminate hostility. I hope to have shown that the scientists in this unusual family shared many of the characteristics of the rest of mankind. There is a very human side to their story which I have tried to interweave with an account of their scientific achievements. I have also tried to popularize some of the basic work of science. Measuring double stars and parallaxes does not now enjoy the glamour of observing black holes and quasars, but to the scientists of 150 years ago, these routine jobs of today were every bit as exciting as our latest fashions are to us --and they were the essential first steps towards our own achievements. I hope to have given non-scientists some understanding of how astronomers first became able to measure the vast distances of the universe. Finally, I hope that this account of the achievements of a man who, as a youth, was so nearly drafted into Napoleon's army, might encourage members of a generation who find it so much more difficult than I did to launch themselves into a rewarding career in astronomy.

Only with reluctance have I referred to the Struve astronomers, throughout this book, by their Christian names. I imagine that in life they set some store by correct and formal modes of address and I myself dislike adopting so easy a familiarity with men whose scientific achievements are far superior to my own. The constant repetition of the name "Struve" which would otherwise have been required, however, would have been both monotonous and confusing. Moreover, years of reading their writings, both published and in letters, and meeting their descendants, have helped me to come to feel that I know these men and can even regard them as friends.

Indeed, one of the unexpected rewards of this work has been meeting the descendants of my two subjects. Apart from a few brief meetings with the second Otto --before my interest in his family history was fully roused-- I have met two other great-grandsons of Wilhelm Struve: the late Professor Gleb Struve, of Berkeley, California, and Dr. Nils Lindhagen of Angleholm, Sweden. I have also met two great-great grandsons: Dr. Wilfried Struve of Karlsruhe, Germany, and Mr. Andrew Struve, son of Gleb Struve. In addition, I have corresponded with Dipl.-Ing. Eugen Struve of Braunschweig, Germany, a third great-grandson of Wilhelm. Each of the first three named, and their wives, have welcomed me and my wife into their homes and provided me with valuable photographs and documents. I owe a particular debt to Dr. Lindhagen who has generously placed both his knowledge as a member of the family and his expertise as an art historian at my disposal. Without his help, this book would have very many fewer illustrations. Still more, his own enthusiasm and encouragement for my project (and his occasional friendly criticism) have done more than he knows to ensure the successful completion of my task.

It gives me pleasure that this book will appear close to several significant anniversaries: the 200th anniversary of Herschel's recognition of the binary nature of double stars (between 1782 and 1803), of Bessel's birth (1784), Fraunhofer's (1787) and Wilhelm's own (1793); the 150th anniversary of the first successful measurement of stellar parallax (1838-9) and of the founding of Pulkovo Observatory (1839), and the 100th anniversary of the undertaking of the *Carte du ciel* (1887).

I wish to thank also the staffs and directors of the various libraries where I have found material to use. Especially I thank Miss Janet Dudley, former librarian and archivist at the Royal Greenwich Observatory, Herstmonceux, for help during and since one all-too-brief visit; Mrs. Enid Lake and Mrs. Pamela Towlson, respectively former Librarian and current Assistant Librarian of the Royal Astronomical Society in London, as well as the (to me) anonymous but no less helpful staffs of the Royal Society Library, the Library of Congress Manuscript Reading Room (Washington, D.C.) and the Hamburger Staatsarchiv. Formal acknowledgements to those who have permitted the use of copyright material appear elsewhere. Dr. Helen Sawyer Hogg read an early draft of the manuscript and encouraged me to persevere, as did Mr. Kevin Krisciunas, who also made available to me his copies of Otto Struve's letters to Simon Newcomb.

I have used source materials in German, French, Russian and Latin (as well as English!). Translations are in general my own, unless clearly ascribed to another in the bibliographical notes at the end of each chapter. I made none from Russian, however, and I am grateful to Professor A.A. Barrett of the Department of Classics, University of British Columbia, for correcting my Latin and to Professor M.L. Hadley, of the Department of Germanic Studies, University of Victoria, for help with some of the more difficult German passages. Readers should remember, however, that many of the letters quoted were written in English, and neither Wilhelm nor Otto completely mastered English spelling, orthography and idiom. I have not put "*sic*" after every mistake.

Help in obtaining illustrations was also given by Prof. Dr. W. Seggewiss, Dr. W.D. Heintz, Dr. P.K. Seidelmann and the late Academician A.A. Mikhailov. Mr. David Duncan prepared many of the illustrations for their final reproduction and drew the maps and line diagrams. I wish to thank the publisher's reader Dr. R. van Gent, whose eagle eye picked up several copying errors in my quotations and who made a number of helpful suggestions. Mr. Gert Kiers, Editor with Reidel, offered help and advice that did much to turn the manuscript into a published book, and Miss Louise Gaudette of Avalon Data Systems Ltd., Victoria prepared the camera-ready pages. Finally, I thank my wife who helped me to read the proofs and accepted the many evenings and week-ends that, of necessity, I spent in the Struves' company rather than hers.

Victoria, British Columbia
April 1st (New Style), 1987

GENERAL NOTE ON DATES, TRANSLITERATIONS, CURRENCIES AND UNITS

Anyone who writes of Russia before the Revolution must deal with the difference between the Julian calendar, abandoned by Catholic Europe in 1582, and by Britain and her possessions in 1752, and the Gregorian calendar, never adopted by Imperial Russia. Throughout the nineteenth century, which happens to cover the entire period during which either Wilhelm Struve or his son Otto were residents of the Russian Empire, the difference between the two calendars was 12 days. Whenever my sources have either given both dates for events in Russia, or at least specified which is meant, I, too, have done so. Wilhelm Struve nearly always headed his letters (or at least those addressed to people outside Russia) with both dates. Otto quite commonly gave only one. In a letter written to Cleveland Abbe in America, dated 13th February 1887, Otto refers to the day being a Sunday, and this makes it clear that the letter was dated by the Gregorian calendar. Given this, and the fact that Otto favoured and worked for the adoption of that calendar by Russia, I surmise that it was his habit to use Gregorian dates when writing to colleagues outside Russia. No guarantee can be offered, however, that he was consistent. Fortunately, there are few matters in this book, if any, for which the uncertainty of 12 days is important. It can safely be assumed that dates of letters written *to* the Struves, are on the Gregorian calendar.

Except in Chapter 3, in which the historical sequence is important, I have avoided breaking up the text by giving birth and death dates for all the people mentioned. I have given this information, so far as I could ascertain it, in the Name Index, where will also be found a brief description of each person whose name appears in the text. The reader who wants to know more than a name should, therefore, turn to the index.

I have found it impossible to be consistent in the rendering of the name "Pulkovo" in Roman letters. In my own text, I have used the transliteration just given --the preferred modern form. The Struves themselves preferred the German form "Pulkowa" and sometimes used "Pulkova" or the French form "Poulkova" and I cannot change these in quotations without doing violence to my sources. Fortunately, there are few other problems of transliteration, since most of the astronomers referred to have German or Scandinavian names. I have not tried to be consistent in the transliteration of such other Russian names as enter the narrative, but have adopted the form of the name I judge to be most familiar --e.g. Uvarov-- unless, again, quotation from a source requires me to use a different spelling --such as Ouvaroff.

The unit of currency in Imperial Russia, as in the Soviet Union today, was the rouble. By the beginning of the nineteenth century, however, paper (or assignat) roubles had depreciated to such an extent that they were worth only about one-third of a silver rouble, and, as a result, there were for a while, virtually two independent internal currencies in Russia. Although it is important to know which was meant in a particular context, sources are not always explicit. Where they are, I have specified the kind of rouble. It is probably safe to assume that, if silver roubles are involved, the fact will be clearly stated. In 1869, when Otto was trying to sell his library (with Cleveland Abbe's help) to American Institutions, he mentioned in a letter to Abbe that U.S.$1 was equivalent to 1.34 Roubles, and that 1 was somewhat over 6 Roubles. This will serve to give some idea of the value of the late nineteenth-century rouble in terms of other contemporary currencies. The reader must make his own comparisons of relative values then and now.

Many units of length will be found in the following pages. I resolved not to bring them all to a consistent system, whether Imperial (British) or metric, partly because the wide variety of units preserves something of a nineteenth-century flavour, but mainly because, again, I could not convert units without doing violence to many quotations from sources. For long distances, of course, the most common Russian unit was the verst. This is, roughly speaking, about two-thirds of an (English statute) mile, or one kilometre. Otto and Wilhelm both sometimes refer to "miles", but by this they mean *German* miles, equivalent to between 4.5 and 5 statute miles (close to 8 kilometres). I found this out the hard way by trying to walk, in limited time, from Elmshorn to Horst, the Struve ancestral home, a distance described by Otto as "one mile". Heights and depths are often given in Wilhelm's surveying reports in French *toises*. These are about six English feet (and nearly two metres); therefore one toise is approximately the same as one fathom. The smaller English units, such as the inch, were legal standards in Imperial Russia. Nevertheless, in the nineteenth century, there were other inches in current use. The *Paris inch* was about seven per cent longer than the English inch (2.54cm). This explains why the Great Refractor at Dorpat is referred to in some sources as the 9.6-inch and in others as the 9-inch. The first is English measure and the second Paris (the one used by Fraunhofer). In addition, a Paris inch was divided into 12 *lines*. A *line*, therefore, was a little less than a tenth of an English inch. In the early chapters, there are some references to temperatures measured on the *Réamur* scale (R). This scale, no longer used, resembles the Celcius scale in representing the freezing point of water by 0°, but it represents the boiling point by 80°. Conversion to Celsius --now familiar even in English-speaking countries-- is therefore simple: increase the Réamur number (whether positive or negative) by a quarter of itself to obtain the corresponding reading on a Celsius thermometer.

For readers without an astronomical background, a few more words of explanation may be helpful. In the eighteenth century, it was customary to refer to telescopes by the *focal length* of their main lens or mirror. Herschel did so almost exclusively --hence we write of his "20-foot" or "40-foot" telescopes. Since, in Herschel's design, the focal length and physical length are almost the same, this gives a good feel for the size of the instrument. Early in the nineteenth century, astromoners began to adopt the still current practice of referring to telescopes by the *diameter* of their main lens or mirror --hence the 9-inch (or 9.6-inch) at Dorpat and the 15-inch of Polkovo. The advantage of this is that the light-grasp of a telescope depends on the square of its diameter. The Pulkovo 15-inch, for example, collected rather more than twice the amount of light that the Dorpat refractor could collect. There is no simple way of comparing the performances of Herschel's and Wilhelm Struve's telescopes (other than looking through each) since mirrors or lenses of the same diameter do not necessarily have the same focal length. Moreover, there are important differences between reflectors (like Herschel's) and refractors (Dorpat). William Herschel's favourite 20-foot had a mirror of about 18-inches diameter. Its light-grasp, ,therefore, was considerably superior to that of the Dorpat refractor, which, however, may well have formed clearer images.

Astronomers also have some special units for their own purposes. Sometimes "astronomical" distances are measured in *astronomical units*, the average distance of the Earth from the Sun and approximately equal to 93 million miles or 150 million kilometres. More often, the great distances between stars are conveyed graphically by the use of *light-years*. A *light-year* is the distance travelled by light in one year (about 6 million million miles, or nearly 10 million million kilometres). It was used as a unit by both Wilhelm Struve and Bessel, but, nowadays, it is usually found only in popular texts. The apparent separation of two objects as seen on the sky is measured by the angle between lines drawn to each from the observer's eye. In the context of double stars and parallaxes, the most important measure of angle is the *second of arc* (1") which is 1/3,600th of a degree. This is approximately the angle between lines drawn to the observer's eye from diametrically opposite points on a penny at a distance of 2.5 miles. Astronomers refer to the position of objects in the sky by coordinates called' *right ascension* and *declination*, which are analogous to longitude and latitude (respectively) on the surface of the Earth. Right ascension is usually given in hours, minutes and seconds (of time) but can be given in degrees, like declination. The difference between the right ascensions of two objects (in time units) is the length of time that elapses between their successive transits of the same observer's meridian. Wilhelm Struve sometimes gave the difference in longitude between two points on the surface of the Earth in time units --signifying the difference between the local times at those two points.

Finally, a brief explanation of the way in which astronomers measure the apparent brightness of stars may be helpful for some readers. The Greek astronomer Hipparchos is usually credited with recognizing, towards the end of the second century B.C., six grades of brightness among the stars that can be seen without a telescope. Certainly the system is ancient. The brightest are termed stars of the first magnitude and the faintest the sixth. Mainly for physiological reasons, a brighter star *looks* bigger than a faint one, and it was long believed that stars of the first magnitude were bigger than others. Traces of this belief are to be found in Wilhelm Struve's writings, although he himself recognized that it was false. After the invention of the telescope, the magnitude system was expanded to include all the faint stars revealed by the new instrument. It did not reach its fully precise modern form until after Wilhelm's death, but it still remains basically the same system thought to have been created by Hipparchos. The non-astronomical reader need remember only two things: (i) the smaller the number, the brighter the star appears to be (a second-magnitude star looks brighter to us than a fourth); (ii) equal magnitude differences correspond to equal ratios of brightness (a sixth-magnitude star is 100 times fainter than a first, an 11th magnitude star is 100 times fainter again).

BIBLIOGRAPHICAL NOTES

Both Wilhelm and Otto Struve used long --sometimes very long-- titles for their major works, and I have used abbreviated titles for a number of them in the references at the end of each chapter.

F.G.W. Struve, Catalogus Novus: Catalogus Novus Stellarum Duplicium et Multiplicium maxima ex parte in specula Universitatis Caesareae Dorpatensis per magnum telescopium achromaticum Fraunhoferi detectarum. Auctore F.G.W. Struve Speculae Dorpatensis Directore. Dorpati MDCCCXXVII (1827). Typis J.C. Schuenmanni, Typographi Academicae.

F.G.W. Struve, Mensurae Micrometricae: Stellarum Duplicium et Multiplicium Mensurae Micrometricae per magnum Fraunhoferi tubum annis a 1824 ad 1837 in specula Dorpatensi institutae. Adjectis est synopsis observationum de stellis compositae annis 1814 ad 1824 per minora instrumenta perfectarum. Auctore F.G.W. Struve, etc. etc. Editae jussu et expensis Academicae Scientarum Caesareae Petropolitanae. Petropoli, ex typographia Academica 1837.

F.G.W. Struve, Description: Description de l'observatoire astronomique central de Poulkova par F.G.W. Struve. St.-Pétersbourg 1845. Imprimerie de l'académie des sciences.

F.G.W. Struve, Etudes: Etudes d'astronomie stellaire. Sur la voie lactée et sur la distance des étoiles fixés. Rapport fait à son excellence M. le Comte Ouvaroff, par F.G.W. Struve, Directeur de l'observatoire central de Russie et membre de l'Académie St.-Pétersbourg, Imprimerie de l'Académie des sciences 1847.

F.G.W. Struve, Positiones mediae: Stellarum fixarum imprimis duplicium et multiplicium positiones mediae pro epocha 1830.0 deductae ex observationibus meridianis annis 1822 ad 1843 in specula Dorpatensi instutis. Auctore F.G.W. Struve. Petropoli 1852.

F.G.W. Struve, Arc du Méridien: Arc du Méridien de 25° 20' entre la Danube et la mer glaciale, mésuré depuis 1816 jusqu'en 1855 sous la direction de C. de Tenner, Chr. Hansteen, N.H. Selander, F.G.W. Struve; ouvrage composé sur les différents matériaux et rédigé par F.G.W. Struve. Publié par l'Académie des Sciences de St.-Pétersbourg (en 3 tomes) 1860. (The third volume is a volume of plates.)

O.W. Struve, Erinnerung: Wilhelm Struve. Zur Erinnerung an den Vater den Geschwistern dargebracht von Otto Struve. Druck der G. Braun'schen Hofbuchdruckerie, Karlsruhe, 1895. (Otto was nearly eighty years old when he wrote this, and deprived of the ready access to a large library that he had enjoyed all his life. Not surprisingly, there are errors of detail (especially in dates) and inconsistencies. Used critically, however, the book is trustworthy for the outline of Wilhelm's life and work. In particular, the account given in it of Wilhelm's illness agrees well with the letters Otto wrote to Airy at the time. I have adopted the convention of citing all works by the elder Otto as by "O.W. Struve". Together with the date, this should avoid any confusion with the younger Otto --who was only eight years old when his grandfather died.)

Three other works are also referred to by abbreviated titles

A von Oettingen, <u>Gedächtnissrede</u> Gedächtnissrede zur Feier des hundertjährigen Geburtstages von Wilhelm Struve (Vierteljahrsschrift der Astronomischen Gesellschaft, Bd 29, pp 67-90, 1894) (As might be expected of such an oration delivered in the late nineteenth century, it is flowery, --almost fulsome-- in tone, but it does give extracts from official university documents that might not otherwise be available) W Henop, <u>Jacob Struve</u> Die Gymnasialdirektor Jacob Struve und die Seinen (Niederelbingen Altonaische Heimatbucher Nr 5, Altona, 1931) (A copy of this may be found in the Hamburger Staatsarchiv It is clearly derived from the <u>Erinnerung</u>, several incidents being related in almost the same words that Otto used It does, however, give useful information about Jacob Struve not found in the <u>Erinnerung</u>)

Olga Tomaschek, <u>Ahnenlisten</u> Ahnenlisten der Astronomenfamilie Struve prepared by Olga Tomaschek (1960) This, too, was printed privately, I owe my copy to the kindness of the late Prof Gleb Struve

The major collections of letters used are the Airy Archives at the Royal Greenwich Observatory, Herstmonceux, the Cleveland Abbe and Simon Newcomb Papers at the Library of Congress, and the archives of the Royal Astronomical Society I have also benefitted from the collection of letters in the hands of Dr Nils Lindhagen

ACKNOWLEDGEMENTS

Full bibliographic references are given at the end of each chapter, and the source of many of the illustrations is acknowledged in the captions. Except for the family tree, line diagrams and maps, illustrations without credit lines were provided by, or with the help of, Dr. Nils Lindhagen. Permission to use previously published and copyrighted material, or letters in archives, has been granted by the publishers and institutions acknowledged here. Thanks are expressed to: the Royal Greenwich Observatory, Herstmonceux, the Royal Astronomical Society, the Royal Society, the Royal Society of Edinburgh, the Astronomical Society of the Pacific, the Astronomische Gesellschaft, the editors of the Astronomische Nachrichten, Sky Publishing Corporation (Sky and Telescope), the Akademie Verlag, Berlin (D.D.R.), the McGraw-Hill Publishing Company, Domus Galileana, the Universitäts-Sternwarte, Bonn, the Universitäts-Sternwarte, Göttingen, the Deutches Museum, Munchen, the Wilhelm-Pieck Universität, Rostock, Pulkovo Observatory, the Library of Tartu State University, the U.S. Naval Observatory, the Dominion Astrophysical Observatory, Swarthmore College and Mr. P.G.E. Corvam. The writings of Cleveland Abbe, held by the Library of Congress are dedicated to the public. At the request of the Royal Astronomical Society, the following file information is provided of pictures reproduced from their archives: James Bradley (Add.188/5), Herschel's telescope (Collected Scientific Papers), Sir James South (Add.94.54), H.C. Schumacher (Add. 188/31), G.B. Airy (PP.7), Sir J.F.W. Herschel (PP 4b), G.V. Schiaparelli (Add. 91/3/22), Sir D. Gill (Add. 91/3/2), W.H.M. Christie (PP 26) and Hermann Struve (Add. 91/3/89). The three pictures of the present day Tartu Observatory were taken and provided by the late J. Toomsoo. The Archives of the U.S. National Academy of Sciences have permitted reference to unpublished biographical information about the second Otto Struve.

CHRONOLOGICAL SUMMARY OF THE LIVES OF
WILHELM AND OTTO STRUVE

Dates given in this summary are Gregorian (New Style)
as far as can be ascertained.

1793 April 15:	Wilhelm born in Altona
1808 July 7:	Wilhelm leaves for Dorpat; arrives on August 10 and matriculates the following day.
1809 Spring:	Wilhelm becomes a tutor in the Berg household.
1810:	Towards end of the year Wilhelm completes requirements for his degree in philology and turns to the study of mathematics, physics and astronomy early in 1811.
1812:	Napoleon's invasion of Russia.
1813 October 30:	Wilhelm receives master's and doctor's degrees simultaneously and, a month later, is appointed extraordinary professor of astronomy at Dorpat.
1815:	Wilhelm marries Emilie Wall in Altona (June 23) about the time of Napoleon's final defeat at Waterloo. After unsuccessfully applying for a position at Mannheim, Wilhelm takes Emilie back to Dorpat (October).
1816 - 1818:	The Livland Survey. First son (Gustav) born in 1816.
1819 April 14:	Otto born in Dorpat, the third son of Wilhelm and Emilie.
1820:	Wilhelm formally appointed professor of astronomy and Director of the University Observatory (Dorpat). Work begins on the arc of the meridian. The Fraunhofer refractor is ordered by the University.
1824 November 10:	Fraunhofer refractor arrives in Dorpat.
1825 February 11 to 1827 February 11:	The census of double stars.
1827 June:	Publication of the *Catalogus Novus*, followed by the award of the Gold Medal of the Royal Astronomical Society.

1830 December: Audience with Nicholas I, who proposes the foundation of
                      Pulkovo.

1834 February 1:   Death of Emilie about a month after giving birth to a
                      daughter (also Emilie) and the death of her oldest
                      surviving son, Alfred. Otto now becomes oldest son and
                      accompanies his father in June to western Europe to order
                      instruments for the new observatory.

1835 February 25:   Wilhelm marries Johanna Bartels. Later in the year, Otto
                      enrols at Dorpat University. Throughout 1835 and 1836,
                      Wilhelm measures relative positions of Vega and its
                      companion, in first attempt to determine the parallax of
                      Vega.

1837:   Publication of *Mensurae Micrometricae*, followed by new
                      measures of Vega (1837-8).

1839:   Wilhelm leaves Dorpat and is made Emeritus Professor. He
                      arrives in Pulkovo April 19 and the new observatory is
                      formally dedicated on August 19.

1840:   Publication of the final value for the parallax of Vega in
                      the *Astronomische Nachrichten* (January).

1841:   Death of Jacob, Wilhelm's father (April 2). Otto's work on
                      precession published. Otto marries Emilie Dyrssen.

1845:   Otto appointed Vice-Director of Pulkovo.

1847:   Death of Maria Emerentia, Wilhelm's mother (July 14).
                      Airy visits Pulkovo. Wilhelm publishes *Etudes d'astronomie
                      stellaire*.

1850:   Otto receives Gold Medal of Royal Astronomical Society
                      for his work on precession.

1852:   Wilhelm publishes *Positiones Mediae*.

1855 October 3:   Otto's son Hermann born.

1857 Spring: First volume of *Arc du Méridien* sent to press.

1858 January 26:   Wilhelm taken ill. Otto's son Ludwig born (November 1).

1862:   Wilhelm resigns from the Academy and as Director of
                      Pulkovo. Otto succeeds him (January).

1863 October: Wilhelm celebrates 50th anniversary of his doctorate.

1864 August: 25th anniversary celebrations at Pulkovo.

1864 November 25:   Wilhelm dies in St. Petersburg. Otto on sick leave during winter of 1864-5.

1867 August 28:   Johanna Bartels-Struve dies.

1868 September 2:   ˮOtto's wife Emilie dies.

1871 August: Otto marries Emma Jankowsky

1874 December 8-9:   Transit of Venus.

1879 August: Otto's first visit to U.S.A. with Hermann, to order lens for 30-inch refractor.

1882 December 6:   Second transit of Venus.

1883 March: Otto's second visit to U.S.A. with Hermann; 30-inch objective accepted.

1884 August-September: Installation of 30-inch refractor.

1887 April: *Carte-du-Ciel* conference in Paris; Otto offers resignation as Director of Pulkovo shortly after his return home.

1889 August: 50th anniversary celebrations at Pulkovo Observatory. Otto confirms resignation shortly afterwards.

1893 April 15:   Centenary of Wilhelm's birth; memorial lecture at Dorpat. Otto moves to Karlsruhe with Emma and Eva.

1894 Christmas: Bredichin resigns as Director of Pulkovo and Backlund succeeds him early in 1895.

1895 May 1:   Hermann becomes director of Königsberg Observatory. Later that year Otto publishes the *Erinnerung*. Ludwig becomes Director of Kharkov at about this time.

1897 August 12:   Birth of Ludwig's eldest son, the second Otto Struve. Award of Damoiseau prize to Hermann in the same year.

1902 March: Death of Otto's second wife, Emma.

1903:   Award of Gold Medal of Royal Astronomical Society to Hermann.

1904:   Hermann appointed Director of Berlin Observatory.

1905 April 14:   Death of Otto.

CHAPTER 1

FAMILY ORIGINS AND WILHELM'S CHILDHOOD

The Struve family has produced many remarkable men and women, and its descendants are still to be found in many parts of the world, but its chief claim to fame is the contributions that its members in five successive generations have made to the study of astronomy. The family stemmed from Holstein, which with neighbouring Schleswig forms the neck of land north of Hamburg that joins Denmark to Germany. These two countries have disputed ownership of the region, in which both German-speaking and Danish-speaking people are to be found. Holstein, even more than Schleswig, has always been culturally Germanic and was regarded as a part of the Holy Roman Empire; but in 1460, Christian of Oldenburg, King of Denmark, was elected Duke of Schleswig and Count of Holstein, thus strengthening the bonds between these two regions and their northern neighbour, bonds that were not without importance in the history of the Struves.

These Struves were small-holders in the village of Horst, some sixty kilometres north of Hamburg. The earliest members of the family who can be identified with certainty were Johann Struve (c. 1710-1778) and his wife Abel Strüven, whom he married in 1743. A gap in the baptismal register of Horst prevents the line being traced farther back with certainty, although Marquard Struve (fl. 1565)[1], a freeholder in the nearby village of Sommerland may have been an ancestor. Part of the difficulty is that the name "Struve", together with variants such as "Strube" and "Strüven" is very common throughout northern Germany, and especially in Holstein. It is probably a Low-German word, related to the High-German *sträuben*[2] (to bristle). A dictionary of the Holstein dialect[3] lists several words (e.g. *struuf, strufig*) that have similar connotations, so the name seems to have begun as a description, either literal or metaphorical, of its bearers. Perhaps its frequency in Holstein was an indication of the character of the industrious, but not very prosperous farmers found there; at least, some traces of such a character lingered in the famous astronomers descended from Johann.

This Johann inherited a small-holding near Horst, and supplemented his income by working as a builder. His wife bore him four children: two died without known descendants (one in infancy). The eldest, Hinrich, left a son and a daughter. He was a sailor and was believed to have been lost on a voyage to the West Indies. The youngest was Jacob, born in 1755. He began the scholarly tradition in the family and all its well-known members are his descendants.

Jacob was to have become a farmer like his ancestors. His mother died when he was only seven, and two years later he was sent to work on a

1

neighbour's farm each summer until he was fourteen. This might have set the pattern for his whole life if he had not been a physically weak youngster: a small growth on his back is said to be attributable to the exertion of those years[4]. Even at that time, however, he attended school in winter. He was an apt pupil, particularly in mathematics --a subject emphasized in Holstein-- of which he remained fond for the rest of his life. His ability quickly became known and neighbouring farmers came to him for help in their calculations of rent and similar matters.

The village schoolmaster was also the village organist and began to teach Jacob how to play the Church organ. The pupil was soon able to substitute occasionally for the teacher, giving Johann Struve the idea that his son should be educated for the much-respected position of village organist. In order that he might learn bass and pedal work, Jacob was sent to school in the small town of Elmshorn, some six or seven kilometres away, which --even today-- is a small market town, still beyond the urban sprawl of Hamburg. The sexton in Elmshorn had a good reputation as an organist and was also Rector of the parish school. Recognizing Jacob's ability, he allowed the lad to join the Latin class. Johann Struve at first objected to his son being taught what, for a farmer's boy, was "a load of useless rubbish",[5] but he withdrew his objections when he learned that Jacob's progress was attracting attention. Pastor Basmer[6] was willing to take the lad into his own home, nominally as a servant, but in fact to help and encourage him to prepare for entrance to the Gymnasium. It took Jacob eighteen months to complete the course of the parish school and, at the end of it, he took the entrance examination for the Gymnasium at Altona, and entered the *selecta*, or academic class, at Michaelmas 1771, when he was not quite sixteen years old.

This Gymnasium, the Christianeum, was founded in 1738-44, during the reign of King Christian VI of Denmark, as part of a deliberate policy to improve the culture of what had been the fortress towns of the frontier regions of the Duchy (as Holstein had now become). Except for the University of Kiel, the Christianeum was probably the most important educational institution in Holstein. It was a typical academic (and Germanic) gymnasium, and the *selecta* was a two-year course of preparation for university entrance. Altona, then completely separate from Hamburg, was too far from Horst for Jacob to continue living at home: he had to find lodgings and some means of supporting himself. Altona was one of the important ports of Holstein, with many opportunities to learn other languages and a great demand for teachers of them. Jacob had scarcely heard of languages other than German and Latin, but he set out to master both English and French so that he could teach them to others and thus earn his living. In little more than a year, he became a particularly sought-after teacher of those languages, but at the cost of spending three years in the *selecta*, rather than the customary two.

On completion of his course at the Christianeum, Jacob chose to go to the University of Göttingen --rather than to the local university at Kiel-- attracted by the fame of Christian Gottlob Heyne, a wide-ranging classical scholar who had gone to Göttingen in 1763 as professor of rhetoric, director of the philological seminar and librarian (later head-librarian). His reorganization of the university library, together with his pioneering efforts in the scientific

Jacob Struve. Original portrait by Kreymann was destroyed during the Second World War.

study of classical mythology, are still remembered today. Jacob intended to study theology and philology (classics), but Heyne soon persuaded him to concentrate on the latter. Once again, Jacob's ability was speedily recognized, and Heyne, himself the son of a poor linen-weaver, well understood that Jacob must earn a living while studying. He took the young student into his own home, and employed him in connection with the philological seminar. Jacob also owed his first appointments after graduation to Heyne's influence. These were positions in various gymnasia in northern Germany, beginning at Harburg in 1780 and culminating in a professorship at his own old school, the Christianeum. in 1791. Jacob stayed in Altona for the rest of his life, becoming the Rector in 1794, retiring in 1827, and living there until his death in 1841.

In Harburg, Jacob met his future wife --Maria Emerentia Wiese, the daughter of a local pastor. Maria, too, could trace her family well back: her father's ancestors lived in Münden (Lower Saxony) in the fifteenth century and her mother's (probably) in Brunswick in the sixteenth[7]. Pastor Wiese himself had spent some years in Russia as court chaplain to the Duke of Holstein-Gottorp, grandson of Peter the Great, who reigned briefly and catastrophically as Tsar Peter III. Like his wife, the future Catherine the Great, Peter had had

Pastor Wiese. Engraving by C. Fritzsch, 1766.

to renounce Lutheranism and to embrace Orthodoxy. Unlike her, he never lost his childhood attachment to his German ways and religion. He neither liked nor was liked by his Russian subjects, and surrounded himself with Holstein soldiers, whom he provided with Lutheran chaplains. Pastor Wiese was one of those and lived with the princely household at Oranienbaum before Peter's accession to the throne. The pastor's wife gave birth to their first child, a girl, there, and Catherine was the baby's godmother[8], although, unfortunately, the baby died. When Peter succeeded the Empress Elizabeth in 1762, he quickly alienated his subjects, not least by attempts to Lutheranize the Russian church --forcing the priests to shave their beards and to wear simple black gowns. There is no evidence that Pastor Wiese was involved in these activities, or even that he approved of them, but it is understandable that he found it expedient

Maria Emerentia Struve. Original believed to have been a Daguerreotype made in Altona in the early 1840s

to return home with his family after Catherine's usurpation of power and Peter's murder before the end of 1762.

Maria Emerentia was born after this return to Germany; her mother died while Maria was still a child. In 1780, at sixteen years of age, Maria met Jacob Struve, and Pastor Wiese himself married the couple in Steinbeck Church on the 10th of January 1783. The marriage lasted happily for 58 years --until Jacob's death. Maria survived her husband by six years. There were fourteen children, of whom seven died in infancy. Five boys grew to maturity and each had a prominent career, although Wilhelm the astronomer was the only one to live on into old age. Two girls made successful marriages: the elder, Christiane (just two years older than Wilhelm) married a Dutch military surgeon, one of whose grandsons (Dr. Wilhelm Henop) wrote a family history, the source of much of our information about Jacob Struve's household. The oldest of the sons who grew to maturity was Karl (b. 1785), a teacher and writer, and an important influence in his brother Wilhelm's early life.

Given Jacob's and Maria's backgrounds, we should expect them to have created a home in which intellectual achievement was encouraged. So also were athletic pursuits. Jacob's physical weakness seems to have been outgrown. His motto might well have been *mens sana in corpore sano*, but he appears to have chosen a slightly different one. His grandson Otto quotes from a letter of his: "*A teneris adsuescere multum est.* We Struves cannot live happily without unceasing work, since from earliest youth we have been persuaded that it is the best and most useful spice of human life".[9] All Jacob's sons, Wilhelm perhaps most of all, lived in accordance with this precept.

Jacob's official duties were to teach classical languages, dogmatics and exegesis, but he never lost his early love of mathematics. Diaries that he kept were primarily for jotting down theorems or problems as he thought of them. Friends teased him for his absent-mindedness when he was preoccupied with mathematics. He often presented problems and demonstrations in the school year-books, several of which are still extant.[10] Despite archaic spelling, their German is simple and clear. He liked practical problems, and illustrated theorems in probability from the then popular card-game of whist, or in statistics from figures of annual enrolments at the Christianeum; but he was interested in more abstract branches of mathematics --such as the theory of numbers-- as well. One of his surviving articles (he called them *Mathematische Kleinigkeiten*) is about "super-perfect numbers" --numbers whose factors, including unity, add up to more than the number itself-- and he also wrote about "friendly numbers" (pairs of numbers, each of whose factors including unity add up to the other, e.g. 220 and 284). The term "friendly numbers" goes back to classical times, perhaps to Pythagoras, so Jacob could combine his loves of mathematics and classics in their study. Moreover, interest in them had recently been revived by no less a mathematician than Leonhard Euler (1707-1783). Jacob's mathematical work was not just a private diversion: he wrote a textbook renowned for its clarity, and received an honorary doctorate from Kiel University in 1813, in recognition of his mathematical studies. He once even wrote to Gauss, pointing out an error in the latter's recently published *Theoria Motus Corporum Coelestium* ... Gauss acknowledged in a letter to the astronomer Olbers that he had indeed overlooked this error when reading the proofs and that it was pointed out to him by "Professor Struve, a schoolteacher from Altona".[11]

Friedrich Georg Wilhelm Struve, the first of the three children his parents were to bring into the world in Altona, was born on 15th April 1793. We do not know much about his childhood. His father's mathematical diaries also contained occasional notes about one or other of the children, but Wilhelm's own son, Otto, remarked that it was striking how seldom Wilhelm's name occurred in them. The household life of the family was simple and regulated: each hour had its allotted task. The principal recreations were

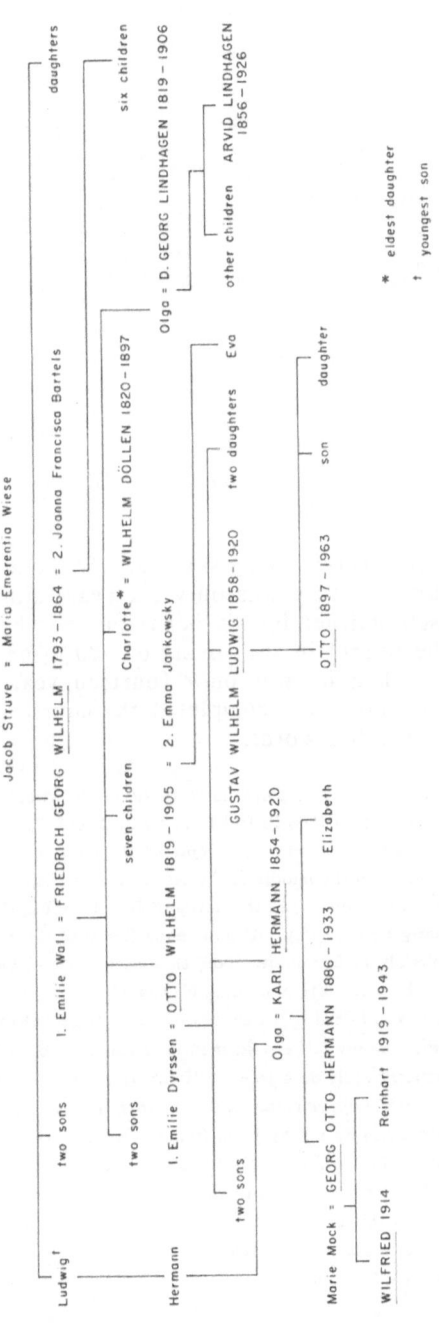

Simplified family tree of the Struves. Astronomers' names are in capitals. The usual convention of placing older children to the left has been abandoned to bring out other relationships.

The Christianeum (from a booklet about that Gymnasium). The Rector's apartments were in the central building on the first floor.

walking, gymnastic exercises and (in winter) skating.   Wilhelm entered the lowest class of the Christianeum when he was only six years old.   He used to say that he distinguished himself neither by his behaviour nor by his position in class; even so, he entered the *selecta* in the spring of 1807 --nearly 36 years after his father had done so-- when he was only fourteen years old, but he spent only one year in the class and never completed the course.   The reasons for this are perhaps best given in Otto's words:

> In this class...Father remained only one year, since he is listed in the Album of the Christianeum : the institute MENSE JUNIO 1808 PRIVATUM VALEDIXIT DORPATUM PROFECTURUS, in contrast to his brothers who all completed the class in due order. This early departure from the institute in the middle of the semester, is perhaps partly to be attributed to the warlike state of the time, after the battle of Jena [October 1806] had made studying in Germany appear generally risky. Also in connection with this there is a remark in Grandfather's diary already in September 1806, that he had asked his oldest son, Karl, living in Dorpat, about the possibility of sending his son Wilhelm to Dorpat. It appears, however, that the decision was forced by an event that strengthened the parents' doubts about keeping him at home. As my father told it in later times, he once fell into the hands of French recruiting officers while on a walk to Hamburg in the suburb of St. Pauli, and was locked up by them in the upper storey of a two-storey house. As a skilled gymnast, however, he escaped from them by a risky jump from the window, and fled back to Altona where the French could not follow him, since Altona was considered to belong to Denmark, and Denmark, as a neutral state, tolerated no foreign soldiers within it. It may seem striking that no confirmation of this incident is to be found in Grandfather's diary; the omission, however, has a natural explanation in that the diary shows a complete gap for that time, namely from autumn 1807 to June 1808.

On the 7th July (25th June) 1808, Father left Altona furnished by the Senior Professor Klausen with a very favourable testimonial on the success of his studies, in the form of a commission from the college of professors, as well as a Danish passport. The latter circumstance had the consequence that later in the Dorpat ALBUM ACADEMICUM he was represented as a Dane, although the Duchy of Schleswig-Holstein belonged to the German Empire and was subordinated to the Danish Royal Family only through a personal union. He reached Dorpat on 1808 August 10 (July 29) after an uninterrupted 35-day journey made troublesome particularly by the bad roads in the recently war-devastated Kingdom of Prussia. As his matriculation shows, the scarcely 15-year old youngster was admitted as a student on the immediately following day. Presumably, therefore, the Altona testimonial of his stay, even though incomplete, in the SELECTA was accepted as sufficient TESTIMONIUM MATURITATIS, or perhaps, at that time, no special entrance examination was required of foreigners.[12]

The story of Wilhelm's capture by a French press-gang is the only one of his childhood to have survived, and even it cannot be dated precisely. Danish neutrality, an essential part of the account, ended on 31st October 1807, as a result of the bombardment of Copenhagen and the seizure of the Danish fleet by the British. Wilhelm's adventure. therefore, could not have occurred any later than that. This alone rules out a much more colourful account of the same incident published by the younger Otto Struve (grandson of the first) that has Wilhelm jumping from his prison into the Elbe and swimming to a ship that just happened to be sailing for Riga.[13] Indeed, it is clear from the elder Otto's account that little had been left to chance in sending Wilhelm to Dorpat. The Napoleonic wars hit Holstein hard, even before the end of Danish neutrality, since her ports were closed to her natural trading partner, Britain. Jacob Struve himself, soon after 1800 lamented the bad effects on enrolments at the Christianeum of the "evil times".[14] He had several of his own children to feed, as well as his brother Hinrich's two, and the chance of sending one away must have seemed attractive to him, as well as being to Wilhelm's advantage. In 1806, when we are told that Jacob first thought of the plan, the journey would have been too dangerous. From then until the battle of Eeylau (February 1807), Napoleon's Prussian campaign would have made Wilhelm's journey very risky. After the Treaty of Tilsit was signed with Russia in July of that year, however, the situation changed. Prussia was subdued and under French domination. Russia was for a while allied with France (and, indeed, later declared war on Britain on the pretext of the Copenhagen affair). Even though this peace with France remained unpopular in Russia, and was never very secure, it at least allowed a breathing space in which journeys such as Wilhelm's were possible. Moreover, at the time that he left home, Napoleon's attention had just been distracted by the beginning of the Peninsular campaign, and the bulk of the Grande Armée was far away. Wilhelm's parents could.hardly have chosen a better time to send him to Dorpat. Equipping him with a Danish passport probably helped to assure him of sympathetic treatment from officials: even many who opposed Napoleon thought the British raid on Copenhagen to have been outrageous.

The Struve family were no admirers of Napoleon: in their family chronicles his victories are usually seen as causes for anxiety, his defeats as occasions for celebration. For them, therefore, Dorpat (modern Tartu, in Estonia) must have seemed an attractive haven. If, as many expected, France and Russia should resume their conflict, Dorpat was well to the north of Napoleon's most probable invasion route. Moreover, there was already a nucleus of the family there. Karl, who had studied at Göttingen University, had gone to Dorpat to teach in the gymnasium, and later became a *privat-dozent* in the new university. He had done this at the suggestion of A.C. Gaspari who, like Jacob, had been one of Heyne's favourite pupils. Gaspari married a cousin of Jacob's wife who had been brought up in Pastor Wiese's household. In 1803, Gaspari was himself appointed professor of history, geography and statistics in Dorpat,[15] and Maria's sister Anna went to live in the Gaspari home there (and later in Königsberg). Thus, the fifteen-year-old Wilhelm was by no means left to fend entirely for himself.

Another person who seems to have influenced the course of events at this point was Heinrich Christian Schumacher. He had been a student at the Christianeum, and he, too, had gone to Livland to be a private tutor, after studying in Göttingen. Jacob Struve had helped to smooth out some difficulties connected with Schumacher's emigration, and Otto tells us that this was the beginning of a lifelong friendship between Schumacher and Wilhelm.[16] Karl and Schumacher were together in Dorpat for a while, but in 1807 --presumably after the signing of the treaty of Tilsit-- the latter returned to Altona. He was to have taken up an appointment in Copenhagen, but was prevented from doing so by the British raid. In 1808, he went back to Göttingen to study astronomy under Gauss --his interest in this subject had been stimulated in Dorpat, his original field of study having been law. Later, Schumacher was appointed to the Copenhagen Observatory, although he spent most of his life in Altona. He is chiefly remembered for founding the *Astronomische Nachrichten*, a journal that is still published and that, in his lifetime, was rivalled in importance only by the *Monthly Notices of the Royal Astronomical Society*.

There is no record of a meeting between Schumacher and Jacob Struve in 1807, but it is scarcely credible that Schumacher would return to Altona and not visit the Gymnasium Rector, especially since he could bring news of Karl. If he then learned that Wilhelm might be going to Dorpat soon, Schumacher would surely have spent some time with the lad. Wilhelm's first position in Dorpat was as a tutor in the very household in which Schumacher had worked in the same capacity, and this suggests that Wilhelm received some help and advice from the older youth. Schumacher might even have helped to rouse Wilhelm's interest in astronomy, since this, too, manifested itself as soon as the latter arrived in Dorpat. Just in the autumn of 1807, a comet appeared in the sky that was bright enough to be seen with the naked eye. Given the superstitions that then (and even now) surround many comets, many people must have associated this with the bombardment of Copenhagen (just as the even brighter comet of 1811 was thought by many to portend Napoleon's invasion of Russia - see Tolstoy's *War and Peace*[17]). Wilhelm must surely have seen this comet, because Jacob would not have missed the opportunity of pointing out such a phenomenon to him. This coincidence of a bright comet and a friend newly enthusiastic about astronomy might well have been a factor in Wilhelm's eventual choice of a career.

NOTES

1   O Tomaschek, 1960, Ahnenlisten p 7

2   W Henop, 1931, Jacob Struve, p 3

3   Otto Mensing, 1933, Schleswig-Holsteines Wörterbuch, Karl Wachholz, Neumunster, Band 4

4   Allgemeine Deutsche Biographie, 1893, Verlag von Duncker und Humboldt, Leipzig, Band 36, p 687

5   O W Struve, 1895, Erinnerung, p 4

6   Allgemeine Deutsche Biographie, (see ref 4) and W Henop, 1931, Jacob Struve, p 4

7   O Tomaschek, 1960, Ahnenlisten p 7

8   W Henop, 1931, Jacob Struve, p 4

9   O W Struve, 1895, Erinnerung, p 9   The letter is also quoted by W Henop, 1931, Jacob Struve, p 7
     The Latin means it is important to be accustomed from youth

10   Copies of some of Jacob Struve's school year-books are still extant and preserved in the Hamburger
     Staatsarchiv

11   C Schilling ed 1900, Wilhelm Olbers -- Sein leben und Seine Werke, Julius Springer Verlag, Berlin,
     Band 11, Abtheilung 1, p 442 Letter No 225, 1809 December 1 (Reprinted 1976 , Carl Friedrich
     Gauss-Werke, Ergänzungsreihe IV, Briefwechsel C F Gauss-H W M Olbers, Georg Olms Verlag,
     Hildesheim-New York, Band 1)

12   O W Struve, 1895, Erinnerung, pp 9-11

13   Otto Struve, 1942, Sky and Telescope Vol 1 , No 4, p 3

14   W Henop, 1931, Jacob Struve, p 5

15   Allgemeine Deutsche Biographie, 1878, (see ref 4), Band 8, p 394

16   O W Struve, 1895, Erinnerung, p 12

17   L N Tolstoy, 1865-69, War and Peace, Book II, part 5, Chap 22

Map of the Baltic provinces of the Russian empire. Dashed lines show approximate boundaries of present-day republics. Names of cities are given as used in Wilhelm's time. The straight line is approximately the meridian through Dorpat. The distance from Tornea to Königsberg is about 800 miles or 1,200 kilometres.

CHAPTER 2

STUDENT DAYS AT DORPAT UNIVERSITY

Estonia is a small country that has not often been left to its own devices. Russia, Sweden, the Teutonic Knights and even Lithuania (when, united with Poland, it was a great European power) have all taken their turn at dominating it. In modern times, only between the two World Wars has Estonia enjoyed an existence completely separate from any other power. In the nineteenth century it was a part of the Russian empire, as it had been since the time of Peter the Great. Before that, the most important influence, other than the Estonians themselves, was the Teutonic Knights. Founded in the late twelfth century as a crusading order modelled on the Knights Hospitaller, the Teutonic Knights came too late to crusade for long in the Mediterranean, but were promised all the privileges of crusaders if they conquered the Baltic littoral and converted its inhabitants to Catholicism. They were more interested in acquiring wealth and lands for themselves: their government of the conquered regions was corrupt and cruel.[1] Sergei Eisenstein cast them as the villains in his film epic *Alexander Nevsky* (1938). Eventually, many of them settled in Prussia, turning it into a Germanic state.

The native Estonians are related to the Finns and speak a Finno-Ugric language --which marks them off from their otherwise very similar neighbours in Latvia and Lithuania. In the fifth century they built the first fortress at Tarbatu --from which both the modern Estonian name of Tartu and the Germanic name of Dorpat derive. Tarbatu is built on and around Toomemyagi (Toome Hill or Domberg) a relatively small hill overlooking the Emajogi River, which flows from Võrtsjärv to Lake Peipus. In the surrounding low-lying country, even a small hill offered a defensible position, but the Russians, under the Kievan Prince Yaroslav the Wise, captured the city in the eleventh century ,and gave it yet another name, Yuryev, after Yaroslav himself. The Teutonic Knights arrived early in the thirteenth century. They built a cathedral on Toome Hill and made Dorpat an important and influential see. When northern Germany embraced Lutheranism, however, the power of the Knights declined. Russia, Sweden and Poland-Lithuania competed to supplant them. The cathedral fell into ruins.

King Gustavus II Adolphus of Sweden founded the first University of Dorpat in 1632. During Peter the Great's northern war with Sweden, the University was removed to nearby Pernau and it was closed in 1710. Alexander I of Russia refounded the University in 1802. His intention was to provide a university modelled on those of Germany for the entire Russian empire, but especially for the provinces of Livland, Estonia and Kurland (roughly the present Estonian and Latvian republics of the U.S.S.R., but divided somewhat differently). Native Estonians did not reap much benefit from this new

Lithograph of Wilhelm's brother Karl made during the latter's years in Königsberg.

university for some time, however, for the enduring legacy of the Teutonic Knights was the Germanic descent of the ruling and landed classes in that area. Many of the new professors had names that betrayed their Teutonic ancestry. On the other hand, these same people forged links with western Europe, and like the Hanseatic League and and Teutonic Knights before them, created a community of the countries that abut on the Baltic Sea. It was natural for a Holsteiner to look to this new university at a time when those of Germany were closed to him, the more so since once (in the fourteenth century) Estonia had been a Duchy subject to the Danish crown, in much the same way that Schleswig-Holstein had come to be.

So Wilhelm arrived in a virtually new university. Even the buildings, centred on the Toome Hill, were incomplete when he first saw them. The cathedral was being reconstructed as the Library and, nearby, the Observatory (also under construction) must have been one of the first things to catch his eye when he registered as a student and received the matriculation number 371[2]. At first, he stayed with his brother Karl who, two years earlier, had married Wilhelmine Sparwardt, the daughter of a merchant and town councillor from nearby Walk (modern Valka). They already had two children, however, and Wilhelm quickly saw that they could not support him as well. Inspired, perhaps, by his father's example, Wilhelm hoped to earn his living while learning by himself teaching others. Karl had been recommended by Gaspari to Herr von Meiner, to serve as a tutor for Meiner's children, and Karl had met his bride in that household[3]. The position had next been held by Schumacher, and, within a month of Wilhelm's arrival in Dorpat, he too was installed in the Meiner household to teach not only those children, but the children of Count von Berg as well.Wilhelm was also awarded a small bursary from the philological seminar, but shortly afterwards, von Berg invited him to become the sole tutor of his own sons and to move into the Berg household.

Meanwhile, acting in accordance with his father's wishes, Wilhelm registered in classes in philosophy (under Jäsche), philology (with his brother) and, surprisingly, astronomy (J.W.A. Pfaff). Presumably this course was the equivalent of a "science elective" in a modern American university. His enrolment in this course is the earliest hard evidence of Wilhelm's interest in astronomy; although Otto also records that his father had quite early expressed a wish to study science and mathematics.:

> Already in spring 1809, just at the end of his first semester at the University, Father had complained in a letter to Grandfather about his dissatisfaction with the philosophical and philological lectures and the hair-splitting and subtleties in these subjects, and at the same time he expressed the wish to learn mathematics. Although Grandfather himself was a passionate worshipper of mathematics, which he called $\kappa\alpha\tau$ '$\epsilon\xi o\chi\eta\nu$ science[4], he forbade him at that time to make that decision, on the one hand calling attention to the influence of philological studies on the mental and moral development of mankind, and on the other predicting a more secure future for his son as a result of such studies, and he believed this wish to be only the expression of a momentary ill-humour[5].

In those days children took the university courses their parents told them to take --even when they were a long way from home and earning their own living. Wilhelm continued in the humanities. Not only were his immediate results good; he was able to use the classical languages throughout his life, and the then new philosophy of Hegel shaped some of his adult attitudes.

All had not been plain sailing, however. Wilhelm's move into the Berg household in the spring of 1809 solved the problem of earning his living, but the work was demanding. The family lived in Dorpat only in the winter; the rest of the time they were at Sagnitz, some 70 versts away. Thus, Wilhelm could not attend many of his lectures and was, for this reason, deprived of his bursary, even being required to refund 150 roubles already paid to him. For a sixteen-year-old boy, Wilhelm's responsibilities were heavy: he had to spend 34 hours a week teaching the four boys. Of these, two, Gustav (or Astaf) and Max were his special care. They were some years younger than he, and their education was entirely in his hands. The youngest boy, Alexander who was six years old, came under his care only later. The eldest, Friedrich, was almost the same age as Wilhelm himself, and Wilhelm had only to help him to prepare for university entrance. In the late eighteenth century, in Russia, to have a foreign tutor for one's children was something of a status symbol. Indeed, a generation earlier the Empress Elizabeth had found it necessary to define minimum qualifications for such positions[6]. Von Berg probably felt that he had found a bargain.

Like many tutors and governesses, Wilhelm seems to have had a more privileged position in the household than other servants: in particular, von Berg treated him as something of a favourite. He took part in the dances and hunts that were part of the round of a great house, and two foster-daughters helped him in his own studies of French and German literature. Nevertheless, guests of the Bergs, nobles themselves, kept their distance from this commoner who had to be careful not to take advantage of his privileged position. He required, according to his son, "an extraordinary amount of tact, prudence and self-control"[7]. Even Friedrich and Wilhelm did not become intimate until many years later. By then, Wilhelm too was a noble and Friedrich, who left the University to fight Napoleon's 1812 invasion, had become a staff officer to General Pashkevich, suppressor of the Polish revolt of 1830 and (at Austria's request) of the Hungarian rebellion of 1848.

The wonder is that Wilhelm made any progress in his studies at all. "Certainly" wrote Otto, his father's progress "was not the result of a very regular attendance at lectures"[8]. In the young university neither professors nor students placed our modern emphasis on lectures. Even so, Wilhelm did not neglect them completely, and Otto found notes of both Gaspari's and Karl's lectures amongst his father's papers. No doubt both of these helped their young relative. Nevertheless, Wilhelm must have studied hard and long after discharging his other duties; Otto said of his father, at this period, "Both mental and bodily fatigue were unknown concepts to him"[9]. At the end of 1810, Wilhelm passed the candidacy examination in philology with honours *and* won a gold medal for an essay entitled *De studiis criticis et grammaticis apud Alexandrinos*. The essay was to have been printed at the university's expense, but, if it was, no copy survives.

Almost immediately, Wilhelm was offered the post of senior history teacher at the Dorpat Gymnasium. Acceptance would have guaranteed him material security, and Wilhelm found this a tempting offer. He was not sure, however, that he wished to spend his life as a teacher, and he *did* want to study mathematics and physics. Having obeyed his father's wishes, he now felt free to follow his own inclinations.

Wilhelm had made friends with Fritz (more correctly Johann Jakob Friedrich) Parrot, the younger of two sons of Georg Parrot, professor of physics at the University, and for many years its Rector. Fritz later became a medical professor at the University himself, and is particularly remembered as being the first person (in 1829) to climb Mount Ararat. Thus Wilhelm came to the attention of Georg Parrot, who became an important influence in the young man's life. Parrot senior was another immigrant from Western Europe. Born in 1767 in what is now Montbéliard (between Strasbourg and Besançon), France but was then Mömpelgard in Wurttemburg, he went to school in the Karlsakademie in Stuttgart with his fellow townsman the future great biologist Baron Cuvier[10]. Later, as a private tutor in France, Parrot met the astronomer Lalande who arranged the printing of a text-book on mathematics that the former had written. Unfortunately the manuscript was lost in the disorders of the French revolution. Parrot married Wilhelmine Lefort, who belonged to the same Geneva family as Peter the Great's companion. He did not come to Russia himself, however, until after her death in 1794. In Livland, he worked as a private tutor, married again, and became the permanent secretary of the Livland Public Utility and Economic Society. As soon as the new University of

Dorpat Observatory 1828-9 from a water-colour by August Matthias Hagen (1794-1878) now in the possession of Nils Lindhagen.

Dorpat was opened he was offered the chair of physics and shortly afterwards followed the ailing theologian Lorenz Ewers as Rector. Parrot welcomed Tsar Alexander I to the University in 1802, and became his friend and confidant. He left Dorpat in 1826, on his election to the Imperial Academy of Sciences and died in 1852 while on a trip to Finland.

Parrot appears to have played a crucial role in persuading Wilhelm to reject the Gymnasium offer and he secured a stipend for Wilhelm, enabling the latter to devote his time to the study of mathematics and natural science. Wilhelm had to give up his teaching duties in the Berg household, although he continued to live there until 1814. The death of Herr von Berg in 1811 seemed only to strengthen Wilhelm's position there, since the widow reposed considerable trust in him and remained his friend for the rest of her life.

Parrot had also played an important part in establishing the study of astronomy at the new University. The first astronomer at Dorpat had been an amateur, Ernst Cristoph Friedrich Knorre. Like Wilhelm himself, Knorre founded an astronomical dynasty: his son and grandson became astronomers. Knorre's observing records date back to 1795, and are still preserved at the University Library at Tartu[11]. He determined the latitude of Dorpat, and became professor and astronomical observer when the University was founded. He made a survey of Livland in 1803 and 1804, and even attempted to determine the longitude of Dorpat. He was Schumacher's first teacher in astronomy. In 1804, J.W.A. Pfaff came to Dorpat as professor of astronomy: he and Knorre did not get on well. Although Wilhelm had enrolled in Pfaff's astronomy course in 1808, Pfaff left in the following year and can have had little direct influence on the student who was to become his most illustrious successor. He did, however, have a considerable indirect influence on Wilhelm, partly through the students whom he did teach --Schumacher, Karl Williams and Georg Paucker-- who later worked with Wilhelm, partly by beginning the regular publication of Dorpat observations (both in formal reports and --for the comet of 1807-- in the local newspapers) and, most importantly, in stimulating the building of the University Observatory. At first Pfaff had had to make do with the attic of a private house. There was no proper slit in the roof, and one instrument had to be carried down to the basement and to another house 15 metres away before it could be used to observe the northern sky. Pfaff bought more instruments and needed better accommodation. Another local amateur, Andreas Lamberti, offered his private observatory in 1807, and even agreed to build an extension, but this arrangement was to last only to 1811. Even before Pfaff's arrival, Parrot had been aware of the need for a permanent building. As Rector, he had set up a commission in 1804 to study the problem. Its members consulted the German astronomer F.X. von Zach, and studied the observatories in Göttingen, Gotha and Uppsala, but not until Pfaff's arrival was the decision taken to build the observatory on Toome Hill, on the ruins of the bishop's palace. The foundation stone was laid on 1808 May 14 (O.S.), less than three months before Wilhelm arrived in Dorpat, and the building was finished on 1810 December 21 (O.S.).

When Pfaff left Dorpat, Knorre resumed charge of the temporary observatory, but he died in the same month that the permanent building was completed. Ironically, therefore, just at the time that the University's instruments could be suitably housed, there was no-one to look after them, and Parrot himself became responsible. He corresponded with C.F. Gauss, offering him the combined chair of astronomy and mathematics; but Gauss, for a variety of reasons, refused, Eventually, J.S. Huth --the director of the university observatory at Kharkov-- was elected to this dual position. He was a sick man, and found the teaching load in mathematics as much as he could carry. He did little in the observatory, although he did publish observations of the comet of 1811. He also petitioned, in 1812, for funds to complete the granite pillars for the support of the instruments. He considered the Observatory to have been constructed to an inconvenient plan, however, and was also concerned that there was no nearby residence for the Director --as late as 1817 he petitioned for the building of such a house. Despite Pfaff's efforts, the instruments available were neither very many nor very good, and Huth left the most important one --a Dollond transit instrument-- packed in the crates in which it had been delivered.

Wilhelm enrolled in Parrot's physics course and Huth's mathematics course, both in 1811. His impressions of Huth have been described by Otto:

> Father never spoke to me of the regular lectures he heard from him. Huth was already a sick man when he came to Dorpat, and on account of this sickness could only imperfectly fill his obligations as professor and still less his tasks as observer and director of the Observatory. Nevertheless, Father respected him as a clever and kindly man, and learned much from his conversations with him. It may therefore, well be supposed that personal

Copy by another artist of the portrait of C.F. Gauss painted by Jensen for Pulkovo Observatory. This copy is at the Universitäts-Sternwarte Göttingen and is reproduced with permission.

intercourse with Huth essentially contributed to the fact that Father, besides his general mathematical studies, followed　by preference the road of astronomical investigation Apart from such occasional discourses with the Professor, who also gave him quite general assignments for study, Father was left to rely on the best text-books then available (Kästner, Euler, Lalande, Schubert, Bohnenberger) to learn the theorems both of mathematics and astronomy, and, in regard to the art of observation, he was in the strictest sense of the word self-taught [12]

Thus it seems that Huth encouraged Wilhelm to work in the Observatory Although Georg Paucker had been appointed astronomer-observer, he was working hard on his dissertation, which he completed in 1813, and also seems to have been glad to have Wilhelm working there.　The whole arrangement no doubt satisfied Parrot, who had the ultimate responsibility for ensuring that the University's investment in the Observatory was not wasted　He must have welcomed the interest of so talented and industrious a student as Wilhelm, the more so, perhaps, as he recalled Lalande's encouragement of him when he was young.　Wilhelm set himself to unpack and to install the Dollond transit instrument. The annual budget of the Observatory barely sufficed to meet the expenses of heating, lighting and stationery. Wilhelm had to use whatever tools were at hand to carve out the cavities in the granite pillars that would support the bearings of the instrument, and the holes that would permit the illumination of the field of view of the telescope.

Wilhelm still spent his summers in Sagnitz, practising surveying with a Troughton sextant that he had bought out of his own meagre savings. An Estonian astronomer has suggested to me that Wilhelm's interest in surveying was stimulated and encouraged by the Bergs, who wished to know more precisely the extent and boundaries of their estate. While this is quite plausible, Wilhelm was following the example of Knorre, Paucker and Huth himself --indeed of astronomers throughout the world who all, at that time, applied their astronomical knowledge to land surveying. In the summer of 1812, however, Wilhelm's enthusiasm for practising his art created trouble for him. Napoleon, whose exploits form a sort of counterpoint to Wilhelm's early life, was about to move against Russia once more. Wilhelm, with his sextant, was arrested by a Russian patrol near Sagnitz, who took him for a French spy. He lost a week's work being taken 150 versts to Pernau, where he had to appear before a military judge. Fortunately, misunderstandings were soon cleared up, but Wilhelm was forbidden to survey that area during the period that invasion was feared. Another effect of Napoleon's campaign might have been more serious. Wilhelm's younger brother Ludwig had come to Dorpat in 1811 as a medical student. In November 1812, he and two fellow-students, one of whom was K.E. von Baer, later known as the father of modern embryology, were ordered to the military hospital in Riga, where they themselves fell ill with typhoid hospital-fever. Wilhelm hurried there to look after his brother until he was sure that Ludwig was out of danger.

Despite these delays and anxieties, Wilhelm took and passed the qualifying examination for the master's and doctor's degrees. At the same time he presented his thesis *De geographicae speculae Dorpatensis positione*. He had completed the installation of the Dollond transit and used it to determine the longitude and latitude of the Observatory more accurately than either Pfaff or Knorre had done. Defence of the thesis was scheduled for 1813 October 29/17, but on that very day the mail (which only came twice a week) arrived in Dorpat with the news of Napoleon's defeat at the so-called "Battle of the Nations" near Leipzig, This, the first major defeat of Napoleon in central Europe, marked the end of any threat to Russian territory. The thesis defence was postponed so that everyone could celebrate. Fortunately, Wilhelm satisfied his examiners on the following day, and like Paucker a few months earlier, was awarded the master's and doctor's degrees simultaneously.[13,14]

On December 2/November 20, on Huth's recommendation, Wilhelm was appointed extraordinary Professor of mathematics and astronomy at the University. Formal appointment as ordinary Professor and Director of the Observatory could not be made until after Huth died in 1818; but Wilhelm was *de facto* in charge of the University Observatory from this first appointment in 1813 until he left Dorpat in 1839 April to become Director of the new Imperial Observatory at Pulkovo.

NOTES

1. E. Christiansen, 1980, The Northern Crusades -- The Baltic and the Catholic Frontier 1100-1525, Macmillan, London, gives a general account of the Teutonic Knights.
2. A. von Oettingen, 1894, Gedächtnissrede, p. 69.
3. O.W. Struve, 1895, Erinnerung, p. 12.
4. i.e. "the chief".
5. O.W. Struve, 1895, Erinnerung, p. 18.
6. W. Bruce Lincoln, 1981, The Romanovs, The Dial Press, New York, p. 291.
7. O.W. Struve, 1895, Erinnerung, p. 15.
8. O.W. Struve, 1895, Erinnerung, p. 17. (A. von Oettingen, 1894, Gedächtnissrede, p. 70, quotes from the minute that required the return of the bursary.)
9. O.W. Struve, 1895, Erinnerung, p. 17.
10. Allgemeine Deutsche Biographie, 1887, Band 25, p. 184. (See Chap. 1, ref. 4.)
11. G.A. Zhelnin, 1969, Astronomocheskaya Observatoriya Tartyskovo (Derptskovo, Yuryeskovo) Universiteta 1805-1948, Publikatsii Tartyskoi Astrophysicheskoi Observatorii imeni B. Struve, Vol. XXXVII, Tartu, p. 12. Most of the story of the early development of astronomy at Dorpat is taken from the early pages of this source.
12. O.W. Struve, 1895, Erinnerung, p. 20
13. O.W. Struve, 1895, Erinnerung, p. 23. See also reference 11.
14. A. von Oettingen, 1894, Gedächtnissrede, p. 70

Rev James Bradley third Astronomer Royal of England (Reproduced by permission of the Royal Astronomical Society)

CHAPTER 3

ASTRONOMY AT THE BEGINNING OF THE NINETEENTH CENTURY

This chapter is designed to give some background information on the state of astronomy when Wilhelm Struve embarked on his career. It may safely be skipped, without any serious effect on their comprehension of the rest of the book, by any readers who are familiar with its subject matter.

The seventeenth and eighteenth centuries were a time of transition in astronomy. At the beginning of that period, Galileo (1564-1642) had put a new instrument, the telescope, into the hands of astronomers, who then learned to exploit its potential and to improve it so that it could be used to the greatest effect. Ironically, a major factor in this improvement was the demand for small telescopes created by the very Napoleonic wars that had sent Wilhelm from his native land to Dorpat. The instrumental developments of this period led to a change of emphasis in astronomy itself. Until the time of Isaac Newton (1642-1727), the central problem of astronomy was to understand and to predict the apparent motions of the planets. Newton solved that problem in all its essentials, and observational astronomers, at least, were free to turn their attention to the fixed stars.

Of the many astronomers who contributed to these developments, two have a special relevance to the life and work of Wilhelm Struve: James Bradley (1693-1762) the third Astronomer Royal of England and William Herschel (1738-1822). Both men attempted to measure stellar parallax, a measurement with which Wilhelm had a special concern (Chapter 8). Parallax is the small annual change in the apparent direction of a star that is the reflection of the Earth's orbital motion around the Sun.

Failure to detect this change was one of the chief scientific arguments against a heliocentric theory of the solar system; the counter-argument that the stars were too far away for parallax to be measurable, although true, did not, perhaps, seem very convincing at the time. Once astronomers did accept that the Earth does move around the Sun, however, the detection and measurement of stellar parallax became a great challenge, and also the gateway to learning the true nature of the stars, since it would enable their distances and luminosities to be determined.

James Bradley, whose birth and death respectively were each almost exactly a century before Wilhelm's, very nearly succeeded in measuring parallax.[1] Each of these men was among the most accurate and respected observers of his day; Wilhelm eventually acquired Bradley's manuscripts for the library of Pulkovo Observatory. Bradley attempted to determine the parallaxes of bright stars that passed close to his zenith, particularly the star γ Draconis (he was observing from near London). The apparent position of a star near the zenith is hardly affected by atmospheric refraction, and it can be determined with a telescope that need be moved only a little and can, therefore, be fairly rigidly mounted. For both these reasons, Bradley could hope to measure the apparent positions of such stars more accurately, and would thus have a higher chance of detecting their parallax. He did show that he could have detected a parallax as large as 2", and he believed that he could have detected one of 1". He seems to have meant by this the total shift, throughout a year, in the apparent position of a star; the modern convention is to define the parallax as half that shift. Since the largest known parallax (in the modern sense) is about 0"75, Bradley's work did give astronomers a very reasonable idea of how large stellar distances are. He also set a standard of accuracy that his colleagues and successors knew they must surpass, if they were to succeed where he had failed.

Even more important than this achievement,[2] however, was Bradley's discovery of the aberration of light, made while he was looking for parallax. The effect of parallax is to make a star seem a little closer in the sky to the Sun (less than 1"!) than its mean or average position. Bradley found a much larger displacement (up to 20") of the star towards a point 90° away from the Sun. It is said that he realized the significance of this while watching yachts on the River Thames tacking against the wind. Because light travels at a finite speed and the Earth is moving, the telescope must be pointed in a slightly different direction from that in which the star really lies, in order to bring that star into the centre of the field --just as a person walking must hold an umbrella in a different direction from that appropriate for someone standing in the rain. The Danish astronomer Ole Römer (1644-1710) had already shown in 1676 that the speed of light could be estimated from the delays in the times of eclipses of Jupiter's satellites when the Earth is distant from Jupiter, compared with when it is closer. Bradley's discovery enabled the velocity of light to be estimated more accurately, provided the size of the Earth's orbit was known. Alternatively it could be used to estimate the size of the orbit, if the velocity of light was assumed to be known. It was as convincing evidence of the motion of the Earth as the effect that Bradley was actually looking for. It was also to have a direct effect on Wilhelm's work, since one of *his* contributions to astronomy was the accurate measurement of the so-called *constant of aberration* which he undertook at Pulkovo.

If Wilhelm's respect for Bradley must be inferred from his acquisition of the latter's observing records for Pulkovo, his admiration of William Herschel is explicitly avowed in several of his own writings, particularly in the *Etudes d'astronomie stellaire.* Herschel's life overlapped both with Bradley's and Wilhelm's, but by the time that Wilhelm had completed his studies, Herschel was an old man and the two never met. There were many parallels between their lives. Both were born in what is now Germany and emigrated to another country, although Herschel, a Hanoverian, did not change his sovereign when he moved to England. Each founded a scientific family --Herschel's son John Frederick William (1792-1871) being Wilhelm's almost exact contemporary did meet him and correspond with him. Quite early in their acquaintance, John Herschel presented Wilhelm with a complete set of the elder Herschel's publications, annotated in the author's own hand. These papers, too, were eventually deposited by Wilhelm in the library of Pulkovo because he believed

Sir William Herschel. Protrait made by Artaud in 1819. (Reproduced by permission of the Royal Greenwich Observatory.)

that in this way he could "make the most worthy use" of "this precious gift"[3]. Much of Wilhelm's own work was a continuation of lines of investigation begun by the elder Herschel.

Although many people remember Herschel primarily for his discovery in 1781 of the planet Uranus, this was by no means his most important contribution to astronomy. Two areas in particular of Herschel's studies were to influence Wilhelm, namely the discovery of the orbital motion of double stars and the work on the "size and construction of the heavens". Like Bradley, Herschel hoped to measure parallax and he followed a suggestion by Galileo that the best opportunity of doing so was offered by double stars --pairs of stars, often of very different brightness, that are close to each other in the sky. Early in his life, Herschel used as a working hypothesis the idea that all stars are of much the same brightness. The members of such pairs, therefore, he supposed to be at very different distances, and he hoped to measure the parallactic displacement of the brighter, nearer star with respect to the fainter. Telescopes had revealed many such pairs, and some time before Herschel embarked on the systematic compilation of a catalogue of them, John Michell (1724-1793) argued on grounds of probability that not all these pairs could be chance coincidences[4]. Herschel published his first catalogue of double stars in 1782[5], and in 1803 he showed that the components of at least one of those pairs must be moving in orbits around their common centre of mass[6]. He it was who coined the term "binary system". Thus, like Bradley, Herschel failed in his stated purpose but made another discovery of comparable importance.

The development of Herschel's ideas on the size and construction of the heavens will be discussed in Chapter 10, in connection with Wilhelm's own work on that subject. Herschel's method was to make what he called "star gages" by counting the number of stars visible in his telescope, down to the faintest that he could see, in many different directions. His first paper on the subject was published in 1785[7], when he still believed all stars to have nearly the same brightness, and with this assumption, he could use his star gages to estimate the extent of the system of stars in different directions, in terms of the average distance assumed for the brightest and supposedly nearest stars --a unit whose size could be at least roughly estimated from Bradley's work. In this area of study, as well as that of double stars, Wilhelm followed Herschel's pioneer efforts, and he became one of those who succeeded where Herschel and Bradley had failed, in the measurement of stellar parallax.

Wilhelm Struve began his own astronomical work at a time when astronomical instruments were being much improved. Much of his significance in the history of astronomy derives from his particular skill in using the newly available telescopes and accessories. As already mentioned, some of these improvements came about as a result of the Napoleonic wars[8]. The earliest telescopes had suffered from two principal defects --chromatic and spherical aberration. The first is the failure of a lens to bring light of different colours to the same focus, while the second is failure to bring all rays, even of the same colour, to a unique focus. Thus, images of bright objects seen through early telescopes were blurred and falsely coloured. Both effects got worse in telescopes of larger aperture --which astronomers need in order to collect as much light as possible-- but can be minimized for a given aperture if the focal length of the objective lens (and therefore the physical length of the telescope) is made as great as possible. In the late seventeenth century, instruments of up to around 150 feet or 50 metres in length were made. They were very difficult to manage and their mountings were nothing like so massive and rigid as those of today. Teams of men were required to move the telescopes, and even light winds could interrupt observation. All kinds of observing must have been very difficult, and precise measurement was all but impossible. About this time, Newton concluded that chromatic aberration could not be removed from a refracting telescope (in which the objective is a lens) and invented one form of reflecting telescope in which the objective is a mirror, which suffers no chromatic aberration and for which spherical aberration can be minimized by careful figuring.

William Herschel's forty-foot telescope. (Reproduced by permission of the Royal Astonomical Society).

Modern astronomers have adopted Newton's solution and all large telescopes now are reflectors, with mirrors of glass (or ceramic) coated with aluminum In the eighteenth and early nineteenth centuries, however, mirrors were made of an alloy of copper and tin, the reflecting surface being given a high polish. They could crack, tarnish easily or even lose their figure completely. Thus, although they were cheaper and easier to make --Herschel, who had to make his own, used reflectors almost exclusively-- reflecting telescopes had their own problems and improvements in refractors were still sought. Several eighteenth-century artisans, including John Dollond, father of the man who made the transit instrument for Dorpat, eventually found that Newton had been wrong. If an objective is made of two or more lenses of different kinds of glass, combinations can be found in which the chromatic aberration of one is very nearly cancelled by that of the other. It is, however, difficult to make a large lens. The glass blank from which it is ground must be free of all bubbles, striae and (particularly) cracks Opticians learned how to produce such blanks only by trial and error, gradually increasing the diameter of the lens they could make until, late in the nineteenth century, they reached what is still believed to be the practicable limit for an objective, around one metre (40inches). In Wilhelm's day the German optician and physicist Joseph von Fraunhofer (1787-1826) had just learned how to make an objective of about one-quarter that diameter, which he used in the telescope that he provided for Dorpat (Chapter 5).

The optical quality of a telescope is not its only important feature. The instrument must also be rigidly and conveniently mounted. Fraunhofer made important advances in this respect also. He developed what is now called the "German form" of the equatorial mounting. One of the axes about which the telescope moves --the polar axis-- is made parallel to the axis of the Earth, so a star, once in the field, can be followed by turning the telescope only about this axis. Fraunhofer also devised a means of doing this by clockwork. Thus, the observer could have both hands free while being able to observe a given object for as long as he wished. Herschel would surely have envied such freedom; Wilhelm Struve could enjoy it. This application of clockwork by Fraunhofer was one of many ways in which Wilhelm was helped by improvements in clock-making that were made contemporaneously with those in optics.

The final important feature of a modern telescope is the ease and accuracy with which its position (and therefore that of the object to which it is pointing) may be read and recorded. For this purpose, the telescope is equipped with accurately graduated *circles* on each axis. Again, important improvements in the art of making such circles were made during the eighteenth century, particularly by the English craftsmen Bird and Ramsden.

Circles on the Dorpat refractor were adequate for pointing the telescope to any object of known position, but would not normally be used for the accurate determination of positions. This was the work of the transit telescopes, such as the one that Wilhelm installed. A transit telescope is mounted so that it can swing only in the meridian of the observatory, and it can be used to measure exactly when a star crosses that meridian. This provides a measure of the local time, if the position of the star is accurately known, or it provides an accurate measure of the *right ascension* (see the General Note at the beginning of this book) of the star, if the observatory is also equipped with an accurate clock. If a transit telescope also has a good circle by which its elevation can be measured, the *declination* of a star crossing the meridian may also be determined. Such a modified transit instrument is called a *meridian circle*. These instruments became the work-horses of any nineteenth century observatory after Reichenbach made one for Bessel, and all astronomers knew how to use them and understood the need for repeated observations to eliminate the effects of inevitable errors of adjustment and graduation from the results obtained with them. Twentieth-century astronomy is concerned with much more than accurate positional measurement, and transit instruments have declined in relative importance: most modern astronomers, the present writer included, have never used one. Wilhelm Struve, however, made himself a master of the particular kind of practical astronomy based on such instruments.

To measure small differences of position between the two components of a double star, or between a star suspected of showing a parallactic shift and reference stars, an auxiliary instrument is required. The most common was a filar micrometer, placed in the eyepiece of a telescope. Once again, Wilhelm was fortunate in arriving on the scene at a good time. Eyepiece micrometers, in which a moving thread is used to bisect the image of each star in turn and the distance moved by the thread is measured by the turning of a screw, were in use at the end of the seventeenth century, but were much improved during the eighteenth. Herschel used them, but was hampered by the inconvenience of his mountings, which required him to use both hands to move his telescope (except for his very largest instruments, which required assistants to move them). As we have seen, Fraunhofer's design left Wilhelm's hands free; moreover, Fraunhofer provided a greatly improved micrometer with the Dorpat instrument --at no extra cost to the purchaser. This was another factor in Wilhelm's early success, as was recognized by no less a person than Alexander von Humboldt. He had persuaded the King of Prussia to finance an observatory in Berlin, and he wished to equip it with a twin of the Dorpat refractor. Schumacher played a part in the negotiations, and in a letter written to him in 1828, von Humboldt stipulated that "The *micrometer described by Struve* must be included in the cost (as it was for the Dorpat instrument), and, if it is not [already] made, then it must be delivered afterwards, *free of charge*"[9]. (Emphasis is in the original).

NOTES

1. J. Bradley, 1729, Philosophical Transactions of the Royal Society, Vol. 35, pp. 637-661.

2. Several authors, 1962, Quarterly Journal of the Royal Astronomical Society, Vol. 4, pp. 38-61 give a general and accessible account of Bradley's work.

3. F.G.W. Struve, 1847, Etudes d'astronomie stellaire, p. 23.

4. J. Michell, 1767, Philosophical Transactions of the Royal Society, Vol. 57, pp. 234-264.

5. F.W. Herschel, 1782, Philosophical Transactions of the Royal Society, Vol. 72, pp. 112-162.

6. F.W. Herschel, 1803, ibid., Vol. 93, pp. 339-382.

7. F.W. Herschel, 1785, ibid., Vol. 75, pp. 213-266. (All Herschel's papers were reprinted in the two-volume The Scientific Papers of William Herschel, edited by J.L.E. Dreyer, et al. and published in 1912 jointly by the Royal Society and the Royal Astronomical Society.

8. H.C. King, 1955, The History of the Telescope, C. Griffin and Co. (Reprinted 1979, by Dover, New York) p. 161. I have drawn on this book for much of the discussion in the rest of the Chapter.

9. K.-R. Biermann (ed.), 1979, Briefwechsel zwischen Alexander von Humboldt und Heinrich Christian Schumacher, Akademie Verlag, Berlin, p. 34. Quoted with permission.

CHAPTER 4

MEASURING THE EARTH

Wilhelm's years as a professor in Dorpat were probably the most fruitful scientifically, and the most rewarding personally, of his life. Into those twenty-five years he packed more than seems possible for any one man to achieve. He was responsible for courses in both mathematics and astronomy, until 1820 when he gave up teaching the former, and he undertook two research projects. The first was the measurement of the arc of the meridian passing through Dorpat, and the second was the search for double stars, which was the most important task he undertook with the Fraunhofer refractor that he acquired for the University Observatory. Wilhelm worked on these programmes concurrently, but they must necessarily be described consecutively. The geodetic work is the principal theme of this chapter, and the astronomical of the next.

After his appointment as extraordinary professor, Wilhelm returned to Holstein to visit his parents, whom he had not seen for nearly six years. He obtained leave of absence in the summer of 1814 and his brother Ludwig, who had decided to return to Kiel University to complete his studies, travelled with him by ship from Riga to the old Hanseatic port of Lübeck. At home, Wilhelm met and fell in love with Emilie Wall, the daughter (not yet 18 years old) of friends of his parents. Her family were of Huguenot stock (their name was originally "Valles") and they had come to Germany, from the Champagne district, after the Revocation of the Edict of Nantes in 1685. The couple became engaged on August 13th, about six weeks after Wilhelm arrived in Altona. Only two weeks later they were separated again since Wilhelm had to return to Dorpat. According to Otto[1], Wilhelm met several German astronomers during this trip: J.G. Repsold the instrument maker in Hamburg (founder of a firm that remained in the family for three generations), Olbers in Bremen, Gauss and Harding in Göttingen, Schröter in Lilienthal and Lindenau in Gotha. Gauss, however, in his correspondence with Olbers, refers to no visit from Wilhelm before 1820[2], and appears not to have known of him at all, even as late as 1815. Wilhelm's first meeting with Friedrich Wilhelm Bessel was also dated by Otto to this same summer of 1814, when Wilhelm stayed in Königsberg on the way back from Dorpat. Bessel had given up a promising career with a shipping house in Bremen, partly because of the encouragement he had received from Olbers, a leading German astronomer who was a practising physician in the same city. Bessel became the director of Königsberg Observatory and perhaps the most renowned mathematical and practical astronomer of his day. He had many interests in common with Wilhelm, and the two men became firm friends and remained so, despite a certain element of rivalry in their relationship, until Bessel's early death from cancer.

Miniatures of Emilie and Wilhelm made by Kreymann (Altona) at the time of their engagement (1814). Copied by Nils Lindhagen from originals once in the possession of Eva Struve.

Wilhelm travelled to Altona again in 1815 to marry Emilie. This time, Karl and his family travelled with him, for Karl had not seen his parents for ten years, and was just then moving from Dorpat to Königsberg. Ludwig graduated from Kiel a few days before the wedding, to which he came, as did brother Ernst from Flensburg. Shortly before the wedding, news reached Altona of Napoleon's final defeat at Waterloo (June 18), so the whole town was in a festive mood. News also arrived that Wilhelm's friend Schumacher, who had been Director of Mannheim Observatory, had resigned in order to return to Copenhagen. Wilhelm's parents and Emilie's urged him to apply for the vacant position. He obtained letters of recommendation from both Olbers and Bessel and, apparently at Harding's suggestion, travelled to Mannheim where he was encouraged by conversations with Schumacher and the minister of the Grand Duke of Baden, in whose gift the appointment lay. The Grand Duke, however, had already promised it to someone else.

Gauss and Olbers discussed this appointment in their correspondence[3]. Naturally, Gauss had been approached for advice and, equally naturally, thought of his own students. In a letter to Olbers written on June 10th --before Wilhelm, according to Otto, had gone to Mannheim-- Gauss discussed the relative merits of Nicolai and Gerling. Replying on July 15th, Olbers told

Wilhelm Olbers. Portrait painted by R.F.C. Suhrlandt. (Reproduced by permission of the Wilhelm-Pieck-Universität, Rostock, D.D.R.).

Gauss that he had already written on behalf of Struve whom he knew "as no other than a very skilled observer who appears well worth a recommendation". Olbers appears, in this letter, to be defending Wilhelm to Gauss who had made a critical remark about "S". Gauss' letters to Olbers at this time, however, betray no knowledge of Struve and the "S" probably stands for Schumacher. Olbers had, indeed, also recommended Nicolai and Gerling, but he wrote off his own influence as "just about zero" compared with that of Gauss. Nicolai received the post and held it until he died in 1846, but the Grand Duke's first choice, apparently declined the offer.

It is unclear whether Wilhelm strongly wished for this position himself, or whether he was merely deferring to the wishes of his parents and parents-in-law. He probably received more support in Dorpat than he would have done in Germany. There is no record that he ever again sought employment outside the Russian empire --although in 1818 he was offered an appointment at the University of Greifswald, which he declined. At any rate, the Grand Duke's choice decided Wilhelm's fate; in October 1815, he took his bride on the 14-day sea voyage back to Dorpat, and there they settled and raised their family of twelve children, eight of whom survived to maturity.

During the winter of 1815-16, Wilhelm was asked by the Livland Public Utility and Economic Society to make a survey on which a new map of Livland could be based. Parrot had been secretary of this Society, and the offer may well have come through his influence. Wilhelm gave a brief account of this work, many years later, in the Introduction to his report on the measurement of the arc of the meridian (referred to here as *Arc du Méridien*):

This work was carried out from 1816 to 1818, as far as the measurement of the triangles, of several auxiliary bases and the astronomical determination, by latitudes and azimuths, of 18 points situated on the sea-coasts between the mouth of the Duna and the city of Pernau [were concerned], and it was finished in March 1819 by the measurement of a principal base-line on the ice of Werz-Jerw [Võrtsjärv], a base-line whose two ends were on the shore of this basin. The result of this work was the geographical positions of 325 points of the countryside, with the evaluation of the elevation of 280 points above the level of the Baltic. The instrument that I had available for the horizontal angles was a Troughton mirror sextant of 10-inches radius, but then I had determined the errors of [its] divisions with care. For measuring the vertical angles I had had made in Dorpat a horizontal sector suitable for the measurement of angles up to 10°. Although this operation did not have the accuracy, in the horizontal angles, of recent works undertaken since 1789 for the determination of the figure of the Earth, it could, however, bear comparison with former measures of Peru and Lapland, and it could offer, in 1819, an arc of $2^\circ 30'$ of the meridian included between the Church of St Mary Magdalene in Estonia, latitude $50^\circ 58'$ and the tower of Kreutzburg Castle in Lithuania, latitude $56^\circ 28'$.[4]

Writing more than thirty years after that early survey, Wilhelm could not resist adding a footnote about the accuracy of his early work:

The accuracy of the triangulation of 1816 to 1819 has since been attested to in a surprising manner. It gave the distance between the two extremities named equal to 140160.6 toises [169.742 English miles]. The same distance has been found, by the measurement of the arc of the meridian, equal to 140162.6 toises [169.754 miles]. In the same way the earlier operation gave the altitude of the threshhold of the Dorpat Observatory, above the mean level of the Baltic, equal to 34.76 toises [222.3 feet]; the new one 34.98 toises [223.7 feet].

These extracts make quite clear that Wilhelm proceeded by the standard methods of triangulation, using such landmarks as castles and church towers, whenever they were available, and specially designed and located signals when they were not. He needed to measure only one fundamental base-line in an area as small as Livland; in the later work, that covered so much larger an area, he would measure several to prevent the otherwise inevitable accumulation of errors of measurement. The measurement of the base-line on the surface of a frozen lake was an elegant solution to the problem of finding a stretch of level ground, although not an original one.

The map based on this Livland survey was not published until 1839 but, long before that, the work established Wilhelm as an authority on the region. After he had completed the survey, no major public work --such as road building or the draining of swamps-- was undertaken without his advice. Wilhelm had done almost all the work himself: even the horizontal sector he mentions was constructed under his supervision by local joiners and clockmakers, and he did the field-work alone, except for the necessary help in

Map of Europe showing places associated with Wilhelm and Otto Struve. The straight line is approximately the meridian through Dorpat, the length of which -- from Ismailov to Hammerfest, approximately 1,800 miles or 3,000 kilometres-- was measured under Wilhelm's direction. The dashed arc is approximately the parallel of latitude that he proposed to measure from Valentia to Orsk.

arranging the measurements and displaying the signal markers. This help was provided by students, particularly by Karl Knorre (the son of Ernst Knorre), who later became the director of the Nikolayev Observatory, and by Baron Wilhelm Lieven, later Quartermaster-General, responsible for all military cartography in the Russian empire.

Surveys of this kind were a regular part of the work of nineteenth century astronomers. Many national observatories were founded precisely in order to provide centres for such work --whether by exploration and discovery at sea, as in seventeenth-century Britain, or by surveying extensive land masses, as in nineteenth-century Russia and early twentieth-century Canada. Large-scale geodetic surveys were undertaken by the French Academy and the Paris Observatory in the late eighteenth century[6] (partly in conjunction with the definition of the metre) and these provided much of the material for Bessel's study of the figure of the Earth. In the nineteenth century, almost every major country organized large surveying projects. Everest's great Indian survey did not long precede Wilhelm's work. Although there were practical reasons for all this activity, national prestige seems to have been at stake too. These projects were the "space science" --or at least the "big science"-- of their day, even to the involvement of the military. Astronomers and soldiers worked side-by-side, and perhaps the later magnificence of Pulkovo Observatory was an early example of the benefits of "spin-off".

There *was* a scientific justification for this work. Knowledge of the figure of the Earth was needed for dynamical studies of the Earth-Moon system as well as because all our knowledge of stellar distances depends ultimately on how well we know the size and shape of the Earth. Wilhelm, a pioneer in the measurement of stellar distances, knew this very well and he emphasized in the Introduction to the *Arc du Méridien* that a major source of uncertainty in Bessel's computation of the figure of the Earth was the scarcity and poor quality of the data available from northern latitudes. He placed the work in its scientific context in a letter written (in English, while he was in England) to the Royal Astronomical Society in 1830.

> It is acknowledged that the further determination of the dimensions of the earth can only be reached by the measures of considerable arcs on its surface. In this view the Indian measure, already extended to 16 degrees of meridian is the most valuable, still more if we consider the proximity of the equator, which makes it to be the only certain point of comparison for all measures in higher latitudes.
>
> The Russian government some years since ordered the measure of the arc under the meridian of Dorpat observatory. Eight degrees are already finished between 52° and 60°. Now the Emperor has commanded that this measure should be continued as far north as nature will permit; and there is no doubt that it will be extended likewise to the southern boundaries of the Empire. If so there will be measured nearly 25 degrees of the Dorpat meridian, which at some future time may be continued through the confines of Turkey to the extension of 37 degrees between cap [sic] North and the southern end of Candia [Crete][7].

The Russian government did not order this large and expensive project solely out of a disinterested love of pure science, but also in order to know better the extent of its empire. Wilhelm understood this and accepted a responsibility to serve his country in this practical way, in return for the privilege of being free to study astronomy. Charles Piazzi Smyth, then Astronomer Royal for Scotland, visited Wilhelm in Pulkovo in 1859 and published accounts of conversations both with him and with Otto. Although Smyth liked to give the impression that he was quoting *verbatim* speeches that were much longer than we can believe he could have remembered, he did make notes and there seems no reason to doubt that Wilhelm did say something like this (much abridged) extract from Smyth's report:

> ...we must accept those now proved political facts, that science cannot always be at the top of the wheel, in a busy working nation. In a community where all others are toiling for

their daily bread, or their national existence, another cannot be allowed to stand completely apart, unaffected by the general anxieties...

...It is not safe, therefore, for any scientific man to forget the country he belongs to or the duties which that connection devolves upon him; he should, on the contrary, be eminently patriotic and filled with ideas of loyalty, and ought likewise to do something almost daily in his country's service if he would daily be regarded by her with favour in his own line.[8]

A few years later, Otto expressed a similar thought to the American astronomer and meteorologist, Cleveland Abbe:

A community will in general always be sooner disposed to devote the necessary means for a scientific undertaking, if it promises some direct practical use. The idea that "science must be supported for itself, since, where this is done, the greatest practical use develops from it" does not readily impress the general public.[9]

Wilhelm and Otto alike would have been amazed at our distinction between "basic" and "applied" research. They carried out both, and took for granted that a scientist should move from one to the other, as occasion required.

Wilhelm's Livland arc of 2° 30', measured between 1816 and 1818 was, however, hardly the measure "of considerable arcs on [the Earth's] surface" in which he was engaged by 1830. He tells us in the *Arc du Méridien* that even in his student days, when practising surveying on the Berg estate, he had been struck by the fact that the meridian through Dorpat passed through more than 20° unimpeded by peaks of any great height above sea level and that, consequently, no major deflections of the plumb line were to be expected throughout its length. In 1819, after he had completed the Livland survey, Wilhelm made a quite definite proposal to the University Council for the measurement of an arc 3° 35' of the meridian extending from the island of Högland in the Gulf of Finland to the town of Jacobstadt in Curland. The University approved the project and the Chancellor, Prince Lieven, submitted it to Alexander I who made a grant for the purchase of instruments. At much the same time that Wilhelm had been surveying Livland, Colonel (later General) von Tenner had been surveying Lithuania. Each had been unaware of the other's work, but Bristen, Tenner's northern terminus, was close to Jacobstadt, and it became easy to link up the two surveys to make a total arc of just over 8°, once the two men realized their common interest.

Wilhelm and Tenner each thought of an extended arc measurement independently, but Joseph Delisle, a Frenchman who-- in the early eighteenth century-- became the first astronomer in the St Petersburg Academy, appears also to have been the first with this idea. His proposal to measure 22° or 23° of the meridian through St Petersburg was approved by the Empress Anna, and in 1737 he measured a base-line with wooden perches on the frozen surface of the Gulf of Finland --from the island on which Peter the Great had built the port of Kronstadt, to the summer residence of Peterhof. He even began some triangulation in 1739 but then abandoned the project for unknown reasons, his work being forgotten until Otto examined his manuscripts at the Paris Observatory over 100 years later.

Results of other surveys in Lapland had been published in 1738 by the French astronomer Maupertuis and in 1805 by the Swedish astronomer Svanberg. These gave discordant results for the length of a degree in those latitudes, and von Lindenau, director of the Seeberg Observatory (Gotha) proposed in 1814 to the chief of the (Russian) Imperial General Staff, Prince Volkonsky, to measure an arc as far north as possible on the coasts of the White Sea, to resolve the discrepancy (as Wilhelm later did). Volkonsky at first agreed, subject to the Tsar's approval, but when von Lindenau insisted on using exclusively instruments made by Reichenbach, the Prince refused to recommend the project unless all instruments required were made in St Petersburg. Ironically, Wilhelm learned this story from von Lindenau himself when they met in Gotha in 1820 as Wilhelm was returning home from Munich --where he had ordered a Reichenbach meridian circle for Dorpat, to be used in his own geodetic work.

Wilhelm, indeed, soon found that he could carry out his project only with the best instruments. The University already owned a Baumann repeating circle (which allows the observer to eliminate much of the effects of errors of graduation by making repeated measurements of an angle) that Wilhelm had planned to use, but he discovered that whatever merits it may have possessed when it was purchased in 1809, it could no longer be considered of the best quality available, because of the improvements that Reichenbach had made in the manufacture of optical measuring instruments.

Wilhelm's need for new instruments provided an excuse for another journey home to Altona. He and Emilie now had three sons, Gustav (b.1816), Alfred (b.1817) and Otto (b.1819), who had never seen their grandparents. Emilie stayed with her parents for a whole year and her fourth son, Conrad was born in Altona in 1821[10]. For Wilhelm, the trip was a working one. He travelled to Munich (in company with Walbeck from Åbo (Turku) in Finland) to order the instruments, and there had his first opportunity to see the progress that Fraunhofer was making in the manufacture of achromatic lenses, and to inquire under what terms he might buy a telescope for Dorpat. In August, Wilhelm and Walbeck spent a week in Göttingen with Gauss[2]. Then, Wilhelm left Emilie and the children again, to join Schumacher who was surveying Holstein and, that summer, measuring a base-line near Braak --a village some twelve kilometres north-east of Hamburg. This was considered important because it would link German surveys with those of other countries. Gauss took part in the measurements; the survey showed that the observatories of Altona and Göttingen were on almost exactly the same meridian. Moreover, Gauss had devised a new instrument --he called it the "heliotrope"-- and wanted to test it in the field. The principle was simple --a mirror placed on a hill-top would reflect the Sun's light and could be seen at a great distance, and be used to measure angles. Gauss had persuaded Repsold to make a prototype and the two men were testing it while working with Schumacher[11]. Wilhelm and Encke (one of Gauss' former students) were invited to join the party, and Wilhelm was no doubt attracted by the opportunity to get to know Gauss better, and to work with more experienced men. He had to return to Dorpat in October, but stopped in Königsberg on the way, visiting Karl and Bessel. At the observatory he met Bessel's new assistant, F.W.A. Argelander, thus beginning a friendship with another famous practical astronomer that lasted not only until Wilhelm died, but which Otto and Argelander continued until the latter's own death.

To return to the great arc measurement; Wilhelm distinguished four phases in its history[12]:

(i)   Tenner's and Struve's surveys in Lithuania and the Baltic Provinces respectively, covering an arc of total length 8° 2'. This comprised all the work until 1831.

(ii)   Between 1830 and 1844, the arc was continued northwards through Finland to Torneå (Wilhelm used the Swedish name, in modern Finland the town is Tornio) at the head of the Gulf of Bothnia, latitude 65° 50', and preparations were made to continue it south to the Dniester.

(iii)   Between 1844 and the end of 1851, the survey was continued southwards to the Danube and northwards through Scandinavia to the Arctic Ocean.

(iv)   Supplementary work, calculation and reduction, culminating in the publication of *Arc du Méridien* in 1860.

Reichenbach made a universal instrument (that could measure both altitude and azimuth) that was to be used in the first phase and which arrived in Dorpat in June 1821[13]. Wilhelm could not take it into the field until August, since both his university duties and a course in geodesy that he was giving to officers of the Imperial Navy kept him in town. Not until 1826 (when he had the help of Paucker, the former observer at Dorpat) and 1827 could he devote several months to field-work on the survey from Högland to Jacobstadt. Despite an exceptionally cold spell for the time of the year (-13°R) and a heavy snowfall in mid-October 1827, the working party kept on until the final base measurement was made on November 10th. The remaining years of the first period were spent in making the necessary fundamental observations in Dorpat and in preparing the results for publication (1832).

Although Wilhelm came to have overall supervision of the arc measurement, he was personally involved only in the northward extension. To the south, from Lithuania to the Danube, was General Tenner's responsibility. Wilhelm was looking north even before he had completed the Livland survey. There were no political obstacles to the extension of the arc into Finland, most of which was then a Grand Duchy in the Russian empire. As early as 1830, Wilhelm presented a memorandum to the Minister for Public Instruction (Prince Lieven, formerly the Chancellor of the University of Dorpat), who submitted it to the new Tsar. Nicholas I approved and made an annual grant of 3,000 silver roubles, which permitted the hiring of native Finnish assistants. In the letter of August 1830, to the Royal Astronomical Society, which has already been quoted, Wilhelm states that two officers, Captain Rosenius and Lieutenant Oberg, were already travelling in Finland to assess the feasibility of the proposal. They reported favourably, and in subsequent summers Wilhelm directed and took part in the field operations himself. He spent the summer months under canvas and, according to Piazzi Smyth, credited a portable lightning conductor of his own construction with saving his life on at least one occasion when lightning struck near the tent[14]. In 1835 he was joined by an assistant, Woldstedt, who married a daughter of Wilhelm's brother Ernst and was a student of Argelander's (now the director of Helsingfors --or Helsinki-- Observatory). Wilhelm and Woldstedt, later joined by Wilhelm's own student Sabler, together worked the forests and marshes of Finland, sometimes measuring rocks projecting only a few feet above water, that were made measurable only because of rare and large refraction anomalies. They completed the task in 1845.

At the same time, Tenner was surveying the governments (provinces) of Volynia and Podolia to the Dniester --boundary with Bessarabia (Roumania). In 1839, the project gained prestige and support from the Academy of Sciences in St Petersburg, which adopted it as its own --on Wilhelm's persuasion. Further extension, however, depended on the co-operation of non-Russian governments. Although the Tornea extension would link the new survey with the old Lapland surveys by Maupertuis and Svanberg, it became obvious that a modern triangulation of that northern region was required, if the maximum possible arc of the meridian was to be measured. The extension to North Cape required work in Swedish and Norwegian territory. No doubt a Russian request to make such surveys would have aroused much the same suspicions then as it might to-day, so, in 1844, Wilhelm exchanged the role of practical man for that of diplomat, and went to Stockholm to negotiate with the Royal Academy of Sciences there on behalf of the St Petersburg Academy. He was received by the King (Sweden and Norway were united under one monarch while having separate institutions) who agreed to give his support. The following year, a Scandinavian reconnaissance party explored the terrain to decide whether or not the survey was feasible. The only change they recommended was to substitute Fugelnaes, near Hammerfest, as the northern terminus because

Cape North, elevation 122 toises [about 750 feet or 250 metres] above sea level is usually subject in summer to thick fogs which arise continually from the warm current which flows along this coast of Norway[15].

In the *Arc du Méridien* Wilhelm approved this "wise choice" which "in sacrificing only and arc of $\frac{1}{2}°$ had to certain knowledge assured the complete success of the enterprise[15].

Swedish and Norwegian scientists made the triangulation in the summers of 1846 and 1847. Since the surveying season lasted only two months in each year, they must have worked hard. A base-line still had to be measured, however, and astronomical measurements made from points in the network of triangles. Russian scientists provided the instruments and cooperated in this phase of the project. Wilhelm, now turned 50 and head of the new Central Observatory in Pulkovo, did not join in the field-work himself, but delegated it to D. Georg Lindhagen, a diplomatic choice since Lindhagen was a young Swede who had come to Pulkovo to finish his studies. In 1854 he married Wilhelm's daughter Olga and later returned to Sweden, becoming the Permanent Secretary of the Royal Academy of Sciences in Stockholm. He spent two

D Georg Lindhagen   Photograph taken in Stockholm by J. Jaeger

summers in Lapland and his detailed reports for those years (1850 and 1851) are presented as appendices in the *Arc du Méridien*. His attempts to make the necessary observations were often hampered by poor weather --he spent hours in the small temporary observatory in order to use every clear period. The strain of the work can perhaps best be judged by its effect on his assistant in the first year, M. Lysander, who took ill and died on the way home; but there was a lighter side to the work as well. On the spur of the moment they joined an excursion to North Cape to see the midnight Sun. They travelled to Hammerfest on a ship that made this excursion once a year and they

> believed they need not be too scrupulous about this little pleasure cruise, especially considering the loss of time hardly amounted to a couple of days   The journey from Hammerfest to Cape North lasted only 7 hours   On the wonderful evening of July 16 our group, unusually mixed both as to languages and nationalities, climbed the famous Cape from the side facing the sea, and greeted with music and libations from the top of the promontory, the midnight Sun, visible at an elevation of 2-3/4° above the horizon[16]

Lindhagen was disturbed by the poverty of the native Laplanders and the contrasting luxury in which merchants from the south, whose hospitality he enjoyed, lived amongst them:

To be sure, the grape does not grow here, but the juice of the grape --wine-- is found in such abundance that one is tempted to believe that it follows a polar attraction of great intensity. After mentioning this circumstance, in the enumeration of the external conditions of our astronomical work, it is well to add that it never produced either accidental or systematic errors in our observations[17].

While Lindhagen was completing work on the northern terminus of the arc, Tenner had pushed to the southern, Ismailov at the mouth of the Danube. Granite pillars were erected at each end point. The southern one recorded the work with Russian and Latin inscriptions, the northern in Norwegian and Latin. Wilhelm's hope that the measurement might be extended across Asia Minor to Crete was never realized. Persuading the Ottoman and Russian empires to cooperate was beyond even his powers of diplomacy. Those two empires were hostile throughout most of his lifetime and the Crimean War broke out not long after the survey was finished.

In all, 258 triangles and 10 base-lines were measured, and the positions of 13 points determined astronomically. Wilhelm was also anxious to compare the Russian standards of length with those used in other major surveys. A considerable part of the *Arc du Méridien* gives the detailed result of these comparisons. His 1830 letter to the Royal Astronomical Society was primarily a request for an exact copy "by the celebrated Master Troughton"[7] of the standard used in the British geodetic measurements (particularly in India). "This was" Wilhelm wrote "the principal object for which I am send [sic] to London by the Russian government"[7].

The three-volume *Arc du Méridien* sums up most of Wilhelm's work on the measurement of latitudes, since it includes an account of the most important results from the Livland survey, originally published earlier. There was some later work done on the measurement of longitude differences, to be discussed in Chapter 9. There was, however, one other major geodetic project carried out in the Dorpat years, the results of which were published separately; namely the determination of the difference in levels between the Caspian and Black Seas. Earlier work, not surprisingly, had given discordant results. The shortest distance between the two seas is about 800 kilometres --along the range of the Caucasus. Surveying was much more difficult there than in the great northern European plain. Alexander von Humboldt had tried to determine the difference between the levels of the two seas during his 1829 journey in Russia. Wilhelm's old friend Fritz Parrot made several explorations in the area, one in 1811 in company with Moritz von Englehardt, professor of mineralogy in Dorpat, during which they, too, attempted a level measurement between the two seas. Fritz returned to Dorpat in 1820 as professor of physiology and pathology, succeeding his father as professor of physics in 1826[18]. In 1829, together with four students, he made his famous ascent of Mount Ararat and again tackled the levelling problem. There can be little doubt that Wilhelm advised Fritz and was in turn stimulated by him to plan a survey of the whole area that could include precise levelling. Wilhelm drew up the detailed plans, but delegated their execution to his former students and assistants G. Sabler, A. Sawitsch and G. Fuss, who carried out the work in 1836 and 1937. In 1849, Wilhelm published the definitive result: within an uncertainty of little more than an (English) foot, the mean level of the Caspian was then 83.67 feet below that of the Black Sea.

If Wilhelm looked upon his geodetic work mainly as the payment of a debt to society, it was nevertheless a major scientific achievement in itself. Many men would have been well content to have done that much with their lives; but Wilhelm was doing other things *simultaneously*, including the work for which he is best known. Moreover, the 1820s and early 1830s were a time of growing family responsibility --Emilie bore him twelve children, of whom eight survived to maturity. During the 25 years as a professor in Dorpat, he gave 20 different courses, on average six times each[20]. Otto summed up his memories of this period in a letter written in 1886 to Sir David Gill (1843-1914), then H.M. Astronomer at the Cape (of Good Hope) who was combining his own astronomical work with the measurement of an African arc on a meridian close to Wilhelm's own:

> Your letter of 28 Jan, gives a clear insight of [sic] the manifold and grand energy which you display at the Cape.
>
> I am by this reminded of my father, how he used to work in Dorpat during the years 1820-35. At that time he worked simultaneously at the Fundamental determination of Dorpat, determined on the Meridian Circle the position of thousands of Double Stars, carried out the Durchmusterung of the heavens according to Double Stars at the Refractor, measured these micrometrically and was himself the chief worker on the Trigonometrical Survey of the Baltic Provinces. And with all that he found time for lectures at the University, was Pro-rector and in a great measure educated his children. It has always been inexplicable to me how he could do all that simultaneously. In the diversity of tasks and the energy to follow these, of all astronomers known to myself you come next to him. May God give you health and strength to carry these out felicitously[21].

Not all details in this letter agree with what Otto himself wrote a decade later in the *Erinnerung*, but the disagreements are trivial in this context. Wilhelm's own publications testify to his activity during this period of his life and we are reminded of Otto's other remark that "bodily and mental fatigue were unknown concepts" to his father.

NOTES

1. O.W. Struve, 1895, Erinnerung, p. 25.
2. Gauss-Olbers correspondence (see Chap. 1, ref. 11), 1815, letters 305-6.
3. ibid., letter 390.
4. F.G.W. Struve, 1860, Arc du Méridien, Vol. 1 pp. X, XI.
5. F.G.W. Struve, 1860, ibid, p. XI.
6. e.g. S. Débarbat, S. Grillot et J. Lévy, 1984, L'Observatoire de Paris, son Histoire 1667-1963, Observatoire de Paris, pp. 18-22, 30.
7. F.G.W. Struve, 1830, letter of August 24 to the Royal Astronomical Society , written in English from Kensington Observatory. Quoted from the R.A.S. Archives with permission.
8. C.P. Smyth, 1862, Three Cities in Russia (two volumes), Lovell Reeve and Co.. London, Col. 2, pp. 183-4.

9. O.W. Struve, 1869 letter of March 19 to Cleveland Abbe.   Translated from the German original. Library of Congress, Cleveland Abbe Papers, container 4.

10. O.W. Struve, 1895, Erinnerung, p. 30.

11. O.W. Struve, 1895, ibid, pp. 30, 31. See also G.W. Dunnington, 1955, Carl Friedrich Gauss: Titan of Science, Hafner Publishing Co. New York, p. 122.

12. F.G.W. Struve, 1860, Arc du Méridien, Vol. 1 p. IX.

13. F.G.W. Struve, 1860, ibid, p. XII.

14. C.P. Smyth, 1862, Three Cities in Russia (see ref. 8) Vol 2, Chap. 4.

15. F.G.W. Struve, 1860, Arc du Méridien, Vol. 1 p. XXIII.

16. F.G.W. Struve, 1860, ibid, p. LXXXIX.

17. F.G.W. Struve, 1860, ibid, p. XCIV.

18. Allgemeine Deutsche Biographie, 1887, (see Chap. 1, ref. 4), Band 25, p. 186.

19. F.G.W. Struve, 1849, Beschreibung des zur Ermittelung des Höhenunterscheides zwischen dem schwarzen und dem caspischen Meere in den Jahren 1836 und 1837 von G. Fuss, A. Sawitsch und G. Sabler ausgefuhrten Messungen. Zusammengestellt con G. Sabler, und im Auftrage der Akademie herausgegeben von W. Struve, St. Petersburg. The result and its uncertainty are quoted by F.W.A. Argelander, 1866, Vierteljahrsschrift der Austronomischen Gesellschaft, Band 1, pp. 31-52, see p. 48.

20. O.W. Struve, 1886, letter of July 1 to D. Gill. The original is in German but Gill prepared and copied a translation with the aid of his German-speaking clerk. That translation is used here and can be found (with the original) in the archives of the Royal Greenwich Observatory (RGO 15.126 ff. 876-880 and 883-886). Quoted with permission.

Pencil sketch (by Vogel) of Joseph von Fraunhofer   (Reproduced by permission of the Duetsches Museum Munchen)

CHAPTER 5

THE GREAT REFRACTOR

When Wilhelm returned to Dorpat with his bride in 1815, he found the observatory still poorly equipped. The transit telescope that he had installed himself remained the principal instrument and the Observatory House that Huth had asked for had still not been built. Indeed, Huth died in 1818, and never lived in that house, while Wilhelm, his successor --although the formalities of the appointment were not completed until 1820-- could not move in with Emilie and their family until after their third son, Otto, was born in 1819. Wilhelm had to walk through the town and to climb the Domberg every time he wished to visit the Observatory. With his characteristic energy and enthusiasm, however, he made what astronomical observations he could. Even before his marriage he had been observing double stars with the transit instrument and, later, with a Troughton refractor equipped with a micrometer. With these limited means, he became the first to confirm Herschel's conclusion that the two components of the bright star Castor were moving in orbits around their common centre of mass.

Besides the universal instrument mentioned in the previous chapter, Wilhelm had also ordered a Reichenbach meridian circle in 1817[1]. When he travelled to Munich in 1820 to order more instruments for the geodetic survey, he had the opportunity to see both how the construction of this meridian circle was progressing and what advances Fraunhofer was making in the manufacture of large lenses. A Munich industrialist Utzschneider had financed Reichenbach in his early days, and began to manufacture glass of optical quality so that the latter should have an adequate supply. For this purpose, Utzschneider went into partnership with the Swiss optician Pierre Guinand --who already had some reputation for the manufacture of lens blanks. Part of their agreement was that Guinand should train a young man of Utzschneider's choice. The man chosen was Joseph Fraunhofer, but he soon quarrelled with Guinand, who dissolved his partnership with Utzschneider and left Fraunhofer in charge of the optical work in Munich[2]. Fraunhofer is chiefly remembered now for his mapping of the absorption lines in the solar spectrum --still sometimes called by his name-- but his principal claim to fame in his own lifetime was as an instrument-maker. In 1820, Wilhelm learned that Fraunhofer was building a "giant refractor" of nine Paris inches aperture. This would be the largest achromatic refracting telescope made up to that time and Fraunhofer had introduced several novel features into the mechanical design that ensured it would be a major instrument, capable of making great advances in the hands of a competent observer.

Understandably, Wilhelm wanted this instrument as soon as he had seen it. On the way home from Munich, he stopped once again in Königsberg and discussed the matter with Bessel. When he arrived in Dorpat, he had a definite proposal to put before the University authorities. He was prepared to give up a new building planned for the Reichenbach circle, and estimated to cost 16,000 to 20,000 roubles, in order to obtain the new refractor. Parrot had been succeeded by Gustav Ewers who, fortunately, also supported Wilhelm enthusiastically, and wrote to the Chancellor of the University, Prince Lieven, "This instrument, the possession of which would raise our observatory to one of the first in Europe, [we] are offered this opportunity to acquire that perhaps will not return. The telescope costs 16,000 assignat roubles". Ewers suggested that the University should buy the telescope with money it had saved, and that the Observatory should retire the debt over 16 years at 1,000 roubles a year. Ministerial approval of this proposal was received on December 20, 1820[3].

The Reichenbach circle reached Dorpat first, however, arriving there in June 1822. Wilhelm interrupted his field work on the arc measurement to unpack, install and get to know his "splendid instrument". Ewers, present at the unpacking, urged that it was "not the best policy to spare necessary costs" in commissioning the instrument. It took 2,074 roubles and 27 kopecks, and 3 1/2 months of work to bring the circle into operation. Wilhelm was well pleased, however, and considered "that the installation here is, in many respects, superior to those that have hitherto been the best, thanks to patterns studied in Munich, Göttingen and Königsberg".[4]

Over two years later (in September 1824) Utzschneider wrote that the Great Refractor was complete, securely packed and on its way. The price, however, had been increased from 8,000 Rhine Gulders to 10,500. Utzschneider claimed that, even so, the factory would make no profit --but he was prepared to renounce a profit in return for the publicity he expected both the instrument and his sacrifice to bring. In Russian currency the price was increased from 5,000 silver (16,000 assignat) roubles to 6,200. The Refractor, packed in 22 cases, travelled through Magdeburg and Königsberg. Baron Wilhelm Wrangel, a naval lieutenant taking Wilhelm Struve's course in geodesy, was sent with a small patrol to Polangen, where the telescope entered Russian territory. Wrangel was sent to escort the instrument to Dorpat, but he broke a leg somewhere near Walk, and after rather unskilful treatment in that town, had to spend a long convalescence in the Struve house.[5] (A few years later, in 1828, Wilhelm Struve also broke a leg while observing with the Refractor. He is said to have remarked that the telescope was rightly called the "Great Refractor" since it had broken two legs!)

The telescope arrived in Dorpat on November 10th* and was carried through the town in festive procession. Again, Wilhelm lost no time in assembling the instrument: his excitement shines through the account (even in translation) that he sent to Schumacher, on December 31st, 1824, for the *Astronomische Nachrichten*:

> On the 10th November last this immense telescope arrived here, packed up in twenty-two
> boxes, weighing altogether 5000 pounds, Russian weight. On opening the boxes, it was

---

*   According to Wilhelm's contemporary account; Otto Struve and von Oettingen, writing much later, give
         the date November 9th.

found that the land carriage of more than 300 German miles [close to 1,500 English miles], had not produced the smallest injury to the instrument, the parts of which were most excellently secured. All the bolts and stops, for instance, which served to secure the different parts, were lined or covered with velvet; and the most expensive part (the object-glass) occupied a large box by itself; in the centre of which it was so sustained by springs, that even a fall of the box from a considerable height could not have injured it.

Considering the great number of small pieces, the putting together again of the instrument seemed to be no easy task, and the difficulty was increased by the great weight of some of them; and unfortunately the maker had forgotten to send the direction for doing it. However, after some consideration of the parts, and guided by a drawing in my possession, I set to work on the 11th, and was so fortunate as to accomplish the putting up of the instrument by the 15th; and on the 16th (it being a clear morning) I had the satisfaction of having the first look through it at the moon and some double stars.

I stood astonished before this beautiful instrument, undetermined which to admire most, the beauty and elegance of the workmanship in its most minute parts, the propriety of its construction, the ingenious mechanism for moving it, or the incomparable optical power of the telescope and the precision with which objects are defined.

The instrument now stands in a temporary position, in the western apartment of the observatory, where observations may be made for an hour and a half in the vicinity of the meridian to about 45° altitude. Next summer it will be placed in its proper position, in the tower of the observatory, under a rotatory cupola, where it may be used for observations in every point of the heavens.[6]

The sketch that Wilhelm used to help him to assemble the telescope is still extant.[7] Fraunhofer himself described it as "only a perspective view...which represents the view from the side on which the clock is situated".[8] It cannot have been easy to assemble a telescope with only that as a guide. Wilhelm used the first hour of observation to compare the performance of his new instrument with the best portable Troughton refractor that he had had until then. Reporting to the University Council he said:

What a difference is seen there! A mountain peak illuminated on the dark side of the Moon, which offers me nothing remarkable in the Troughton, I recognized, by means of the Giant Refractor as consisting of 6 peaks well separated from each other. One of the most difficult of Herschel's double stars I recognized immediately as belonging to the third class.*[9]

Wilhelm added:

, I believe the telescope can be boldly placed alongside the giant reflecting telescope of a Herschel, for if the latter has a greater light-gathering power, our achromat surpasses by far any reflecting telescope in the precision of the images...By clockwork the telescope can

---

*   It is not clear what, exactly, is meant here. A double star of Herschel's class III would not have been one of his "most difficult". Perhaps Wilhelm saw a faint, distant companion (missed by Herschel) to a much closer double. Fraunhofer said that Bessel had a similar experience with another Munich telescope (p.49).

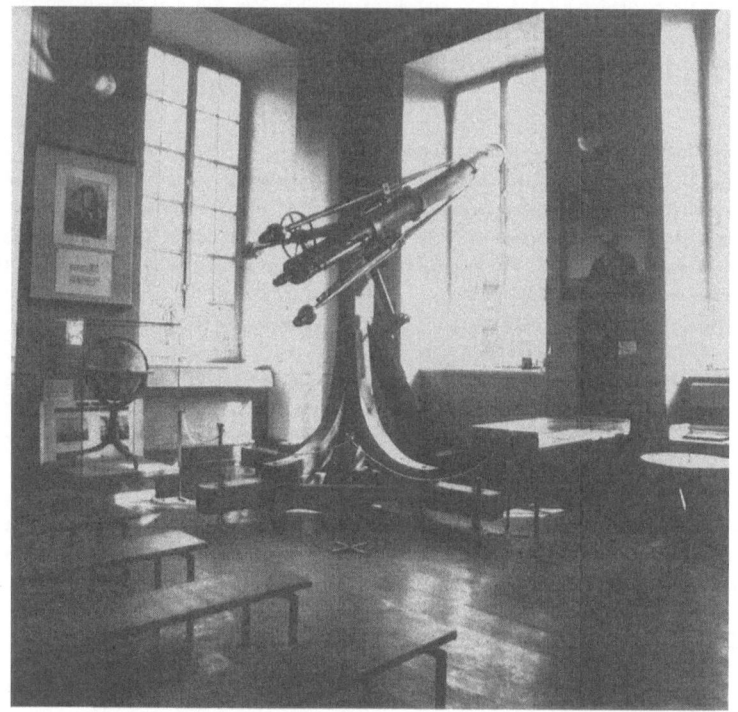

The Great Refractor today (by courtesy J.M. Tomsoo).

be given a uniform motion similar to the velocity of the fixed stars, so that the star remains in the field of view. This is here, notwithstanding that the famous astronomer Bode has now for several years explained that it is impossible...[9]

   This praise for a telescope of less than ten inches aperture may seem exaggerated. Of Herschel's largest telescopes, one had twice the aperture and the other nearly five times. The larger of these was never fully satisfactory, however, and mirrors of speculum metal were decidedly inferior to those of silver (or aluminum) on glass with which we are familiar. Fraunhofer's telescope was the largest achromatic refractor that had been made and its definition probably *was* good enough to compensate, to some extent, for its smaller aperture. The mounting and drive were major advances. Fraunhofer had created, apparently single-handedly, the "German" equatorial mounting used for practically all the great nineteenth-century refractors, and which also influenced some of the twentieth century reflectors, such as the 1.8-m telescope in Victoria, B.C. The clock drive was of particular importance: it helped Wilhelm to use the Great Refractor efficiently, and in time made astronomical photography possible. Fraunhofer's justified pride in his creation is obvious from the account he gave of it in 1824 --published in the *Astronomische Nachrichten*-- to the Royal Bavarian Academy of Sciences. He wrote:

The most critical test of a telescope, as is well known, is the observation of double stars, and by this the effectiveness (i.e. the resolution) of the new achromatic refractors is shown to be much greater than that of reflecting telescopes. Thus, for example, Bessel in Königsberg discovered with an achromatic refractor from here, whose objective has an aperture of 48 lines* that a double star ζ Bootis, classified by Herschel with his reflector as class IV is at the same time also of class I; i.e. he recognized that next to the primary there is a companion star what Herschel did not see. Similarly, many other fixed stars, which were earlier very often observed with [reflecting] telescopes, have been first recognized as double stars, in recent times, with achromatic refractors.[10]

Fraunhofer's design may even have had some effect on the outward appearance of modern observatories. With his "German" equatorial mounting, a telescope could be pointed to any part of the sky above the horizon, otherwise inaccessible regions being reached by reversing the telescope about its polar axis. When the Great Refractor arrived in Dorpat, the central tower of the Observatory was still incomplete. As Wilhelm tells us, he had to use the telescope at an open window, whose frame severely limited the view of the sky. In the winter of 1824-5, Saturn happened to be about as far north in the sky as it can be, and Wilhelm, limited to objects of less than 45° altitude, could not observe it. Small wonder that he wrote:

I wait with impatience for the time when I shall be able to use this instrument in a more favourable position, in order to undertake a new admeasurement of the bodies of our system.[11]

This "more favourable position" required the construction of what Wilhelm called a "rotatory cupola" atop the observatory. Domes are now the distinguishing features of observatories, and small ones are mass produced, but the Dorpat cupola was a forerunner of modern domes and Georg Parrot designed it *ab initio*. In 1825 he had few precedents to copy, although some pre-telescopic observatories (such as Tycho Brahe's) may have had rotating domes.[12] The Great Refractor was dismantled and cleaned in November 1825, and re-erected in its intended home early the following month.

Dorpat University was quick to recognize its debt to Fraunhofer's skill. In June 1825, on the request of the University Council, Tsar Alexander I gave Fraunhofer a diamond ring valued at 3,000 roubles. Wilhelm, too, received a diamond ring when the telescope had been installed. For his part, Fraunhofer completed the micrometers he had designed for the instrument and sent them to Dorpat at no extra cost. Wilhelm estimated their worth at 1,200 roubles.

Fraunhofer's remarks on the use of double stars as a test for his telescope seem almost prophetic, but he probably had a shrewd idea of the plans made by Wilhelm, whose interest in these objects was already well-known --his first catalogue of them appeared in 1820. By the time the Great Refractor arrived, Wilhelm already had had a decade of experience in observing double stars with the transit instrument and the Troughton refractor. His first major project with the new instrument was to make a census of all double stars down to declination 15° south, which resulted --in 1827-- in the publication of the

---

* See general note p. xiii.

Dorpat Observatory in 1860   Copied directly from the colour lithograph made by Louis Höflinger and owned by Nils Lindhagen

Catalogus Novus Stellarum Duplicium et Multiplicium. This catalogue contained 3112 double stars (but two were duplicates), many of which are still known by their running number in it, prefixed by a capital Greek Σ (for Struve). Later observers adopted similar conventions to number their own discoveries, but Wilhelm did not claim to have discovered all the stars in his catalogue. His aim was to list all known double stars. He was careful to give credit to the Herschels (father and son), to South and to Bessel for discoveries they made which he had merely confirmed. Indeed, he included 74 pairs, on the authority of one or other of these observers, without confirming them himself. He claimed 2,508 of the stars listed as his own discoveries (including 141 in the 1820 catalogue) and admitted that even even of these, 165 had been discovered independently by Bessel, South or John Herschel.

In 52 pages of Latin introduction, Wilhelm explained his plan to count the double stars in the sky visible from Dorpat. To complete the survey, however, he had to place some limits on it. Stars more than 15° south of the equator never reached an altitude or more than 16° at Dorpat and Wilhelm found that the atmosphere was rarely stable enough for him to decide whether or not those stars were truly double. He also limited himself to stars brighter than ninth magnitude. Herschel had classified double stars by their apparent separation. The closest pairs were of class I, while components of pairs of class V were separated by at least 30". This classification is no longer used, but Wilhelm adopted it with some modifications. He defined classes I and II more precisely than Herschel had (separations of less than 4" and of 4" to 8" respectively) and set the upper limit for class IV at 32" rather than 30". He did not include more widely separated pairs (Herschel's classes V and VI) in the *Catalogus Novus*. Wilhelm described his method of observing:

> I attacked single zones (of declination) through the hours of right ascension in such a way that, by moving the telescope in declination through the space of a single zone [i e $7\frac{1}{2}°$], individual stars that were conspicuous in the small telescope, the finder, were led into the field of the large telescope to be subjected to examination Such is the ease of motion of our instrument, that by placing the hand on one of the two spheres that keep the telescope balanced, the telescope itself immediately moves precisely in the direction in which the hand presses, no vibration of the complete instrument arising from this Thus, in the space of an hour, more than 400 stars could be examined And the image is so distinct, that, given experience, for the most part it could be decided on the first glance whether or not the star was double [13]

Wilhelm's speed of working has often aroused comment. It was remarkable even if he was giving his best, rather than his average performance One out of approximately every 35 stars examined proved to be double. for these he had to stop, read the circles, record the position ad the magnitudes of the components and estimate the class of double. He must indeed have decided at first glance that most of the other stars *were* single. He soon learned to use only the best nights, when the images were steady enough for quick decisions, but he lamented the long intervals between such nights and the fact that even some that began well often deteriorated quickly. In writing of the local climate, he not only gives some insight into the conditions under which he worked, but discusses a problem that still divides astronomers: should a major telescope be easily accessible, or on a site where it can be used on the largest possible number of nights?

CATALOGUS NOVUS

STELLARUM
DUPLICIUM ET MULTIPLICIUM

MAXIMA EX PARTE

IN   SPECULA

UNIVERSITATIS   CAESAREAE   DORPATENSIS

PER

MAGNUM TELESCOPIUM ACHROMATICUM FRAUNHOFERI
DETECTARUM.

————————

AUCTORE

F. G. W.   STRUVE,
SPECULAE DORPATENSIS DIRECTORE.

DORPATI MDCCCXXVII,
TYPIS J   C SCHÜNMANNI, TYPOGRAPHI ACADEMICI.

Title page of the Catalogus Novus, from a copy in the writer's possession.

I have added this historical exposition of the completed work to bring out a judgement from it about the nature of our climate with respect to observation There are those who believe that a large telescope cannot be applied with success to the contemplation of the sky, except very rarely On the contrary I have found, from the middle of February right up to the end of October, by far the majority of calm nights enjoy such atmospheric conditions that our telescopes can be pointed with success, if not for the whole night, at least for most of the hours especially [immediately] after sunset The months of November, December and January are, indeed, in our climate, for the most part hostile to observation, with the sky very often covered and when the atmosphere, especially as the cold grows, displays images that move intolerably, because --if I am not mistaken-- of the moisture continuously precipitated by the cold Nevertheless, nights of great cold that enjoy the requisite conditions do occur Thus, the last sweeps of 1825 on the four nights of 28th to 31st of December lasted through most of the night The thermometer stood below $-16^{\circ}R$ And in the middle of the month of February 1827, the last examination of the whole effort was made in a temperature of $-19^{\circ}R$, with the sky extraordinarily still But when after the middle of the night the thermometer fell to $-21^{\circ}R$, image motion increased --which made me put an end to observations that had been continued through eight hours [14]

Wilhelm began his observations for the *Catalogus Novus* on 11th February 1825 --just three months after the telescope arrived in Dorpat and while it was still in its temporary location.   He could not use the instrument from 9th November 1825 until the 9th December, when he resumed observation with the telescope mounted under the new "rotatory cupola". He continued on 44 nights until 30th April 1826, when he began field work on the arc measurement Summer nights at a latitude of 59° are, of course, short --many of them too short to be useful for the kind of observations Wilhelm wished to make.   He began observing again on 16th September, and completed his census (except for a re-examination of some zones observed in unfavourable conditions) on the 129th observing night, 11th February 1827.[15]

Wilhelm did most of this work himself. His first regular assistant was Ernst Wilhelm Preuss, a journeyman weaver. Preuss was in hospital in Dorpat, and the hospital director asked Wilhelm if he could find work for this apparently dying man. It soon became obvious that Preuss had considerable talent for both computation and observation.   From 1823 to 1826 he was engaged as an astronomer on the second of Admiral Kotzebue's three voyages of circumnavigation. He returned to Dorpat in 1827, and served as Wilhelm's observing assistant until 1839, when he died.

Wilhelm's quick working precluded accurate determination of the positions of double stars, although he claimed a precision of better than 1' in each coordinate, He could have done even better, but he judged that it would have been a waste of time and effort to do so, since he could easily determine accurate positions with the Reichenbach meridian circle. This he did, and the results appeared in another publication on double stars, the *Positiones Mediae* (1852)

Wilhelm well knew that Herschel had begun to observe double stars in an attempt to measure stellar parallax, and that it was almost by accident that the latter had found that at least some double stars are physically related pairs

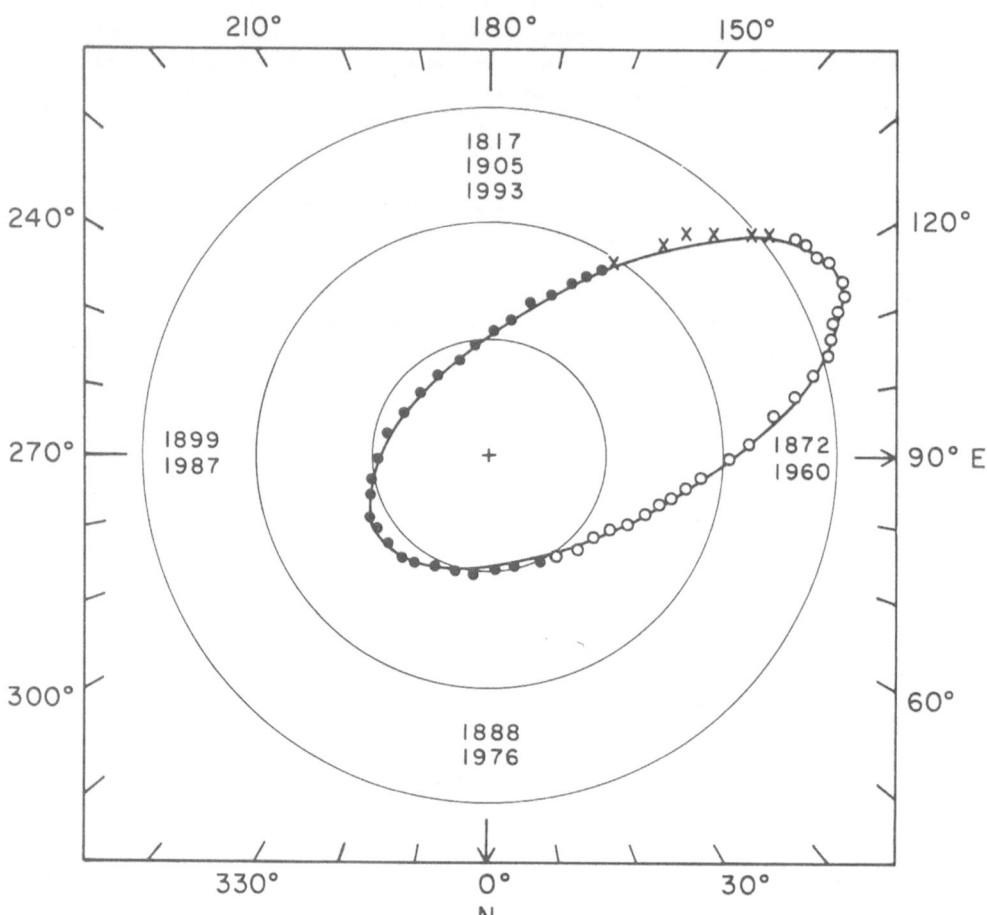

*The orbit of the fainter component of 70 p Ophiuchi around the brighter, as recently computed by M.D. Worth and W.D. Heintz (*Astrophysical Journal *Vol. 193, p. 647, 1974.) North is to the bottom and east to the right. Position angles are marked around the outer edge, and the dashed circles indicate steps of two seconds of arc in the separation. Filled circles are based entirely on Wilhelm's observations, open circles partly on both Otto's and Hermann's. Dates indicate when the fainter component (which takes over 88 years to go round the brighter) is approximately at each of the cardinal points with respect to its companion.*

orbiting around each other. Wilhelm also wished to determine parallax and still believed that Herschel's method --determining the change in position of a bright star relatively to a faint one near it in the sky but assumed to be much farther away-- was the best, provided a suitable "optical" pair could be found. The *Introduction* to the *Catalogus Novus*, therefore, contained a discussion of the

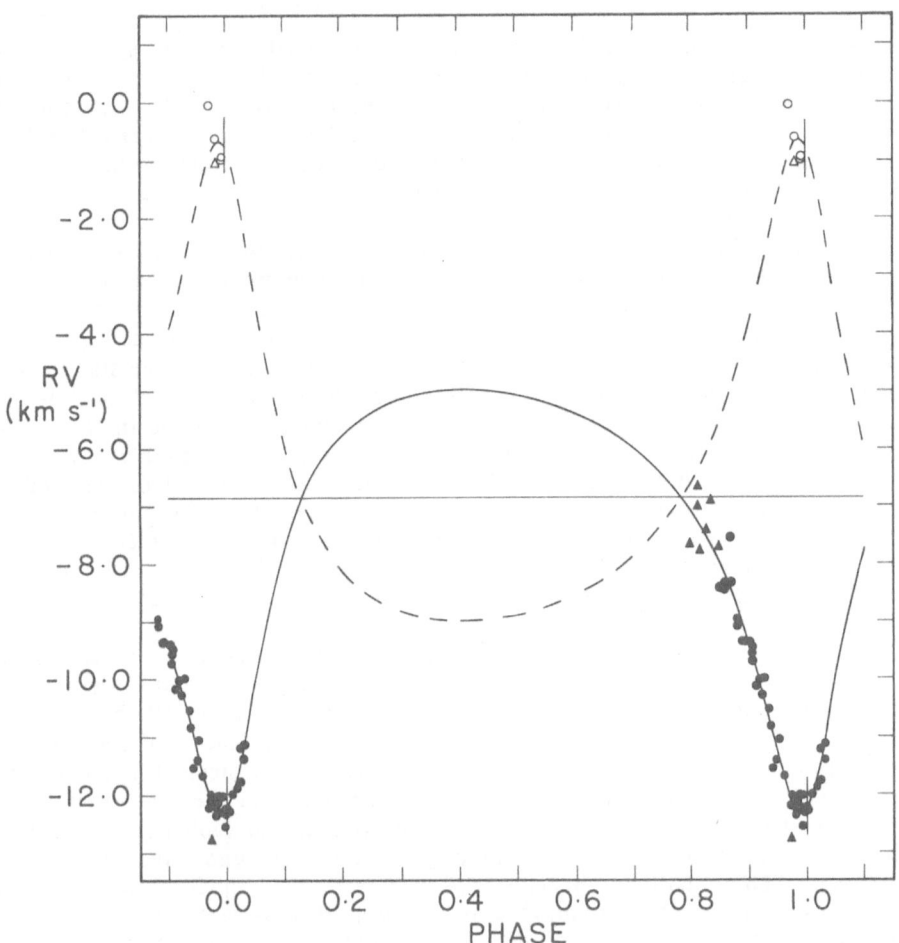

This diagram shows how the line-of-sight velocities of the two components of 70 Ophiuchi change as the stars move around their orbits. The two curved lines show how the velocities are expected to vary (the dashed line refers to the fainter component). A negative velocity means that the star is approaching the Sun. (The system as a whole is approaching the Sun at about 7 kilometres a second.) Individual observations of the brighter star, made at the Dominion Astrophysical Observatory from 1966 to 1986, and a few of the fainter star made in 1982 and 1983 are shown. Each small division on the horizontal axis represents a little less than nine years. The stars had the velocities corresponding to phase 1.0 at the beginning of 1984.

fraction of stars listed likely to be optical pairs. Here Wilhelm began to develop ideas on stellar statistics that he presented much more fully, twenty years later, in his *Etudes d'astronomie stellaire.* In 1827, his ideas were in transition. He knew that his own discoveries showed that stars were not all of the same intrinsic luminosity, nor distributed uniformly in space, yet his attempts to estimate the probability that observed double stars were chance optical combinations were partly based (at least tacitly) on the assumption of both those ideas. For this and other reasons, he tended to over-estimate the proportion of physical pairs among his discoveries, and he believed that scarcely any at all --except the very widely separated ones-- were optical pairs.

First, Wilhelm compared the numbers of pairs in the different classes I, II, III and IV. If most of them were chance combinations of stars at very different distances from us, he argued, the numbers in successive classes should increase as the square of the mean separation of pairs within each class, since the ratios of these squares would be those of the areas around one star in which another would have to fall to form a double of a given class. Wilhelm predicted that the numbers of doubles in four classes would be in the ratio of 1:3:12:48 respectively. He found, however, nearly 1,000 doubles of class I, between 600 and 700 in each of II and III and just over 700 in class IV (for which he admitted that his observations were incomplete). From these results he estimated that 98 per cent of his class I pairs, 90 per cent of class II and 58 per cent of class III were physical pairs. His basic argument is certainly correct, even though we might not agree with the precise figures.

Next, Wilhelm considered the ratio of double to single stars among stars of different apparent brightness. One star in every 12 brighter than seventh magnitude was found to be double, but only one in every 35 brighter than ninth. He argued that if most pairs were chance optical combinations, their frequency would be independent of the apparent brightness of the component stars. The real separation of physically related pairs, on the other hand, would be much the same, whatever their distance from us, so such pairs should look closer together, when viewed from a great distance, and thus be harder to discover. Thus, Wilhelm argued, the decrease in frequency with apparent brightness is more easily explained if we suppose that most pairs are genuine binary systems. This is an argument that depends on the assumption that most stars are of much the same luminosity, but since, *on average*, fainter stars are farther from us, it does lead to an essentially correct conclusion.

Wilhelm then considered the 872 doubles of class I for which he had been unable to determine the magnitude of each component separately. He found that the components of 263 pairs were almost equally bright and those of another 365 differed by only a magnitude or less. The components of very few pairs differed by 1.5 magnitudes or more, and those of only two differed by more than 5 magnitudes. If most pairs were chance associations, he argued, large magnitude differences should be most common. He failed, here, to recognize the importance of what we now call observational selection. It is *much* harder to detect a double whose components are of different brightnesses, and the closer they are the harder it becomes. Thus pairs consisting of equally bright stars are over-represented in the *Catalogus Novus*.

Wilhelm then turned to the proper motions of stars --their transverse motions across the sky relative to the average of all other stars. He credited the Italian astronomer Giuseppe Piazzi with first recognizing that the two components of 61 Cygni have the same proper motion. Wilhelm used catalogues compiled by Bradley, Bessel and Piazzi to pick out other pairs with large proper motions. He argued that if both members of a pair had the same proper motion --especially if it was large-- they were probably related even though no orbital motion could yet be detected. On the other hand, if one member had a large proper motion not shared by the other, the two stars probably formed a chance alignment. These arguments are still considered quite correct and are used today. To them we can add similar arguments based on line-of-sight velocities, which Wilhelm could not measure. By his arguments, he identified δ Equulei as an optical pair --later its brighter component was found to be a genuine binary with a period of five years-- and η Cassiopeiae, 70 p Ophiuchi and 61 Cygni (all quite correctly) as physical binaries. He does not seem to have drawn the conclusion in 1827 that stars with large proper motions were likely to be nearby (he did later, see Chapter 8). Each of the three systems just mentioned has a larger parallax than that of Vega, the star Wilhelm selected as specially suitable for parallax measurement. His friend Bessel did choose 61 Cygni for that purpose and succeeded in determining the first reliable parallax. Had Wilhelm selected any of the three, he might have reached that goal first -- but they were not suitable for the method he adopted.

Wilhelm's last argument was to consider all stars (10,229) down to the seventh magnitude in the part of the sky he had observed. Combining observations from several sources, he was careful to compare his own estimates of magnitudes with those made for the same star by others, so that no large error should be made by mixing magnitude scales set up by different people. Then he tried to calculate the probability that any one of these 10,229 stars should have a chance companion within 4" (class I), 8" (class II), 16" (class III) and 32" (class IV). He concluded that none of the observed double stars, with both members brighter than seventh magnitude, in classes I and II were chance optical combinations and, indeed, only one such pair was to be found in all four classes (which contained 207 pairs). His tacit assumptions about the luminosities and distribution of stars would certainly have affected this result, and we would nowadays use a different statistical analysis based on the Poisson distribution. Poisson's work, however, was not published until some years after the *Catalogus Novus*.

Not content with these arguments from pairs of stars, Wilhelm next discussed the 52 systems he had discovered that contained three or more components. He showed that it was very unlikely that these were chance combinations. Then he drew attention to some very wide pairs of bright stars separated by several minutes of arc. The components were of second, third or fourth magnitude, and Wilhelm believed them to be related in some way. Subsequent research proved him right more often than not. The components of pairs like 16 and 17 Draconis, or ν^1 and ν^2 Draconis are at much the same distance from us and travelling together through space, even though they may not be in orbit around each other. Wilhelm also spotted the connection between the stars in the Orion trapezium, although we now believe they are expanding from a region of common origin, rather than forming a permanent group. He even drew attention to the Southern Cross (which he could never have seen himself) pointing out that one first-magnitude star, two second-magnitude stars and one each of third and fourth would hardly be within a circle of just over 3° radius by chance. Four of these five stars are now considered to belong to the Scorpio-Centaurus group --another group of young stars that probably had a common origin. Surprisingly, he did not comment on the seven stars of second and third magnitudes in the Big Dipper, five of which we know to be travelling together through space.

All these arguments led Wilhelm to believe that optical pairs, suitable for the determination of parallax, should be sought among the widest doubles. Herschel, on the other hand, thought --before he demonstrated the true binary nature of many doubles-- that only pairs with a separation of 5" or less would be suitable for parallax measurement. In the *Catalogus Novus*, Wilhelm identified α Lyrae (Vega) and α Andromedae (separations 42" and over 80" respectively) as pairs that he believed to be optical and specially suitable for measuring the parallax of the brighter component. As we shall see in Chapter 8, he eventually succeeded in measuring the parallax of the former.

The *Catalogus Novus* was the first of a trilogy on double stars and it brought Wilhelm immediate fame. It was published in June 1827; he received the Gold Medal of the Royal Astronomical Society of London that same year. This award has a prestige among astronomers similar to that of the Nobel Prize, and it was given to Wilhelm specifically for his work on double stars. He received another diamond ring from Tsar Nicholas I[16] and was also selected a corresponding member of the Academy of Sciences in St Petersburg. Two years later, Alexander von Humboldt visited Wilhelm in Dorpat. Von Humboldt made clear, in letters home to his brother and to the French astronomer and physicist Arago, that he was fascinated by Wilhelm's discoveries and found Wilhelm one of the more interesting people whom he met in Russia.[17] Thus, before he was forty years old, Wilhelm was recognized as one of the leading astronomers of his day.

Wilhelm's other major works on double stars were the *Mensurae Micrometricae* (1837, discussed in Chapter 8 in connection with stellar parallax) and the *Positiones Mediae* (1852, already mentioned). Both contain observations that are still of value. For example, Wilhelm's 1834 discovery that the brighter component of the pair Σ 2367 in his *Catalogus Novus* is itself a close double is recorded in the first of these. The orbital period of this pair is now known to be about 92 years and recent calculations show that Wilhelm's discovery observation was in error by only 0"02 in separation and 4° in position angle (a measure of the orientation of the line joining the two stars).[18] His observation agrees much more closely with the computed orbit than do many subsequent ones by other observers.

During all the scientific activity recorded in this and the previous chapters, Wilhelm's duties as a university professor and his family life claimed and received his attention. Emilie returned to Dorpat in 1821, giving birth to her fifth son, Heinrich, in July 1822 and her first daughter Charlotte, in May 1824. That same year, the family circle was further enlarged when Ludwig returned to Dorpat as Professor of Medicine. The relationship between Wilhelm and his brother, who had been students and were now professors together, became even closer since Ludwig had married Emilie's sister Conradine. Otto thought the years 1827 and 1828 (when he was eight or nine) the high point in family life.[19] The Wall grandparents spent a whole year in Dorpat, and the

University's silver jubilee was celebrated then. Otto also (mistakenly) places von Humboldt's visit in that period. The happy time ended rather suddenly. Otto himself was seriously ill in February and March of 1828, and Uncle Ludwig, to whom Otto later credited his recovery, died in April after only a short illness. Gustav, Otto's eldest brother, died in November, followed the next month by his barely three-year-old sister Alexandra. Each had succumbed to typhoid. The other children escaped that, but all contracted measles about the same time. Wilhelm had his famous accident with the Great Refractor in August of that year, and broke his right leg. Meanwhile, Emilie bore her eighth child (and sixth son), Bernhard, in 1827 and another daughter --also christened Alexandra-- in 1829.

Otto recalled 1830 as an especially busy year for his father.[20] Wilhelm and Emilie, their seven living children, a nephew and a foster-child Theodor (Wilhelm's brother Ernst's son) and a children's nurse all set out for Altona to visit the four grandparents. They chartered a carriage and left Dorpat in the middle of May, travelling night and day except for rests in the larger cities. The journey itself was one of the few times Wilhelm had leisure to devote to his family. After travelling through Riga, they spent ten days with Karl in Königsberg. In Berlin they stayed with Emilie's relations and, although there were no proper roads, journeyed through the Brandenburg Mark, Mecklenburg, Lauenburg and Holstein to Altona. As usual, Wilhelm made this a working holiday. After two weeks he left the family in Altona and visited Paris and London. With the same gift he had shown in early life for walking into any conveniently nearby wars or revolutions, he arrived in Paris quite literally on the eve of the 1830 revolution. Early in the morning of July 27 (the first of the three days known to the French as "les Trois Glorieuses") he went to the Paris Observatory, but by evening it was unsafe for him to return through the streets of Paris to his hotel. He had to stay at the Observatory until July 30, when its Director, Alexis Bouvard, one of Laplace's former students, the first to study the irregularities in the motion of Uranus that led to the discovery of Neptune and, above all, a staunch republican, guided this loyal servant of the Tsar of all the Russias through the barricades to his hotel.[21]

Otto says that Wilhelm went to England chiefly in order to meet Sir John Herschel and Sir James South --his principal competitors in the observation of double stars. Wilhelm, as we have seen (Chapter 4), said that the main reason the Russian government sent him was to secure an English standard measure. The two statements, of course, are not inconsistent. Certainly Wilhelm began lifelong friendships with South and Herschel on that visit. The elder Herschel had died three years earlier, but this was the occasion on which John Herschel presented a complete set of his father's papers (annotated in William's own hand) to Wilhelm (see Chapter 3). South, at that time, was chairman of a committee to reform the *Nautical Almanac* --the annual volume of astronomical tables then published by the Royal Observatory, Greenwich, and still published jointly by the Royal Observatory and the U.S. Naval Observatory, under the title *Astronomical Almanac*. Wilhelm was at once co-opted to this committee and was specially thanked in its report for his contribution. Here he first met George Airy, later Astronomer Royal and a close friend of both Wilhelm and Otto, and Captain (later Admiral) W.H. Smyth, the father of Charles Piazzi Smyth who visited Pulkovo in 1859 (Chapter 7).

Sir James South. (Reproduced by permission of the Royal Astonomical Society.)

After some weeks, Wilhelm returned to Altona by sea, but immediately took part in an assembly of natural scientists in nearby Hamburg. He was an invited speaker there, and his audience included the Swedish chemist Berzelius, the Danish physicist Oersted and other contemporary scientists of the first rank. At the end of September the family, reunited at last, returned to Dorpat. With both winter and the university session approaching, the return journey was more hurried than the outward one had been. Moreover, Emilie was pregnant again and gave birth in December to her fourth daughter Olga --who, in due course, married the Swedish astronomer D. Georg Lindhagen, whose work on the arc of the meridian is discussed in Chapter 4.

After 1820, when Martin Bartels came to Dorpat as professor of mathematics, Wilhelm was at least relieved of the obligation to give courses in that subject. By then, however, he was teaching surveying to officers of the General Staff of the Imperial Navy. Otto, who had studied under his father, explains that Wilhelm coped with his heavy teaching load by giving informal tuition rather than regular lectures.[22] He could do this especially in astronomy, since not many students enrolled in a course at any one time. He arranged that the professor of astronomy, who had to work at night, should be exempt from the obligation to take a turn as Rector or Dean, but he was several times Deputy Rector. On occasion, he left the telescope to discipline parties of roisterous students (no doubt to their surprise). He also led the University fire brigade. Most of the time this involved only keeping the equipment in good order, but on January 30, 1829 a major fire threatened the University's main building. Because of the extreme cold (-30°R), water supply was a problem, but Wilhelm confined the blaze to a small part of the building. The affair made headlines and brought Wilhelm an expression of thanks from Nicholas I.[23] Wilhelm also served on the University building commission and, after Georg Parrot left in 1827, took over the administration of the Cathedral park, which was just outside his house.

On January 30, 1832, Wilhelm was elected a full member of the Academy of Sciences. This would normally have obliged him to move to St. Petersburg, as Georg Parrot had done five years earlier. The Tsar, in confirming the election, however, allowed Wilhelm the unusual privilege of continuing to live and work in Dorpat. The Academy had an observatory in St Petersburg, but its instruments were so far inferior to those in Dorpat that expecting Wilhelm to move would have made no sense. From Dorpat, he could travel at least to the formal meetings of the Academy.

In this same year, 1832, Sir James South returned Wilhelm's visit. South had planned a large telescope of his own and, in 1829, acquired an object glass of 11-3/4 inches aperture. He commissioned the instrument builder Troughton to build a telescope even bigger than the Great Refractor, but, as soon as it was complete, South objected, alleging in particular faults in the mounting. Troughton was prepared to make good the defects, but South would not let him near the instrument and came to Russia to see if the mounting of the Great Refractor was all it was reputed to be. He went to St Petersburg where Wilhelm was to meet him and to bring him to Dorpat. English visitors were a rarity, however, and South, being both wealthy and eccentric, attracted much attention, especially from the Tsar's younger brother, the Grand Duke Mikhail

Pavlovitch --another eccentric, who, on his wedding night, preferred to inspect a regiment of guards, rather than to stay with his bride, the Grand Duchess Helena. When Wilhelm arrived in the city, the Grand Duke insisted on taking both him and South to Moscow and personally giving them a guided tour of the city. Only after that, was the English astronomer free to spend some weeks in Dorpat. He studied the Great Refractor, but also enjoyed the hunting in the neighbourhood. About this time, Emilie gave birth to her eleventh child -- another son called Gustav, who did not survive infancy. She appears to have recovered in time to make a great impression on her visitor who, Otto says, became her "fervent admirer".[24]

South was elected a corresponding member of the St Petersburg Academy and wrote to Wilhelm in 1833 to express his appreciation. His return to England, however, had renewed the feud with Troughton (and the latter's partner, Simms) which spilled over into the letter:

> Troughton Simm's conduct still continues equally base; and I have resolved to print the correspondence, and when printed I will transmit you a copy -- Simms I suspect is the Leader of the Wickedness, and old Troughton is the victim of his iniquity.[25]

South refused to pay Troughton's bill and this refusal led to legal proceedings late in 1833 which South lost. He smashed the mounting of the telescope, leaving only the object glass. Nine years later, the American Astronomer O.M. Mitchel was shown all that was left. He described how South led him to a large room, littered with the fragments of a great telescope:

> "Here sir" exclaimed Sir James "you behold the wreck of all my hopes. Here I have expended thousands, and flattered myself that I was soon to possess the finest instrument in Europe; but it is all over, and there's an end."
>
> I remarked that the object glass was still in his possession and might yet be mounted so as to realize his hopes and expectations.
>
> "No" said Sir James "Struve has reaped the golden harvest among the double stars, and there is little now for me to hope or expect."[26]

Mitchel suggests that this was the real reason for South's extraordinary behaviour over the telescope. That irascible man had, however, shown himself more than once to be petty and foolish, and it is remarkable that he preserved his friendship with Wilhelm if the latter's priority in the observation of double stars had caused him such chagrin.

Another highlight in the Struve family life was the Golden Wedding anniversary of Jacob and Maria Emerentia on January 10, 1833 (December 29, 1832). The Russian branch of the family could not return to Altona in winter and was, indeed, itself split for the occasion. Wilhelm had to give his inaugural address to the Academy in St Petersburg, and Emilie, Charlotte and Theodor went with him. They stayed there with Joseph Dyrssen, a family friend and Holstein merchant whose daughter (also Emilie) became Otto's wife nine years later. The rest of the family stayed in Dorpat. Aunt Conradine (Ludwig's widow) gave a ball in honour of the occasion and of the opening of her own "educational institute for young ladies". The main family celebration in Altona lasted several days. Jacob, although retired for seven years, was still a notable person in town. Amongst the presents was "a most magnificent drinking cup decorated with the celestial sphere, presented by Wilhelm Struve in the name of the 19 children and grandchildren in Dorpat".[27]

Events took a less happy turn later in 1833. Wilhelm spent the summer as scientific advisor to a chronometer expedition on the Baltic Sea led by General Schubert for the purpose of determining longitudes of points on the Baltic coastline. The expedition went as far as Holstein, so Wilhelm was able to see his parents, but on his return home he found a serious situation. The oldest surviving son, Alfred, had a diseased hip which, despite repeated operations. had grown worse. Emilie, once again pregnant, would scarcely leave his bedside; only a nephew Adolf (Karl's son), who now lived with the family and was a medical student, gave her help and occasional relief. Wilhelm was summoned to St Petersburg in November for a meeting about a proposed new central observatory for the Russian empire. Reluctantly, he went and, Otto tells us:

> Towards Christmas he returned from his journey and found his son in a hopeless condition and his wife expecting a confinement within a few days. On New Year's morning 1834 our sister Emilie was born and on the following morning our brother succumbed to his ten-month illness. To be sure, it seemed that Mother, despite her heavy grief, quickly recovered from the delivery, so that she could leave her bed for a while, but on 20 (18) January she suddenly had a fit of spasms with violent pains in the back and sides, which the physician took for pleurisy. In the course of the following days she apparently improved her situation; however, she felt her end near, and in this belief she called first Father and then me, as eldest son, to her bed, in order to take her leave and say many things that lay on her heart. On 1st February (20th January) it almost appeared as if her feelings had deceived her and she was going to stay with us; however, already around noon-time a new spasm attacked her and in a few minutes her dear life was at an end. The autopsy carried out on the corpse the next day showed that the physician had erred in his diagnosis. All parts of the body were normal and found to be without traces of disorganization. From this it must be concluded that only nervous excitement and enervation, resulting from the unceasing care of her son who departed before [her], was the cause of her death.[28]

Wilhelm's early achievements owed much to the support and love that Emilie had given him. Her organization had enabled the growing family to live in the none-too-spacious Observatory House. Although it had seemed roomy enough at first, Otto says, it became very cramped as more children were born. At least three nephews --Adolf, Theodor and Ludwig's son Hermann-- lived with the family at different times. Loft rooms were added to the house, but before that, Emilie had to enforce strict rules in order to house everybody and still to arrange family parties and celebrations. Wilhelm had a small study in the house, but it was often used as a sick-room, and was a bedroom for one of the boys by night. Fortunately, the children had the run of the Cathedral park. Emilie was also responsible for the education of the children (at least so Otto says in the *Erinnerung*, he gave a different impression in the letter to Gill quoted in Chapter 4). The children did not see much of their father, except at

Wilhelm Struve's house in Dorpat (by courtesy J.M. Tomsoo).

meal times, on occasional walks and during skating and gymnastic practices. The boys were sent to school when they were six or seven years old, and, to earn money for this, Wilhelm gave up his leisure time (!) to teach in school as well as at the University. Emilie appears to have been the dominant formative influence in her children's early years. She was as hard a worker as her husband, *and* she was pregnant for nearly half her married life. She had just turned 37 when she died --worn out. Without her, Wilhelm's story might have been very different.

NOTES

1   F G W Struve, 1860, Arc du Méridien, Vol 1 pp XII

2   H C King, 1955, The History of the Telescope, (see Chap 3, ref 8) gives more details of the relations between Guinand, Utzschneider and Fraunhofer

3   A von Oettingen, 1894, Gedächtnissrede, p 73

4   A von Oettingen, 1894, ibid , (Reproduced with the permission of the Astronomische Gesellschaft)

5   O W Struve, 1895, Erinnerung, p 37

6. F.G.W. Struve, 1826, Memoirs of the Royal Astronomical Society, Vol 2, pp. 93-100. The article is described as "Communicated in a letter from Professor Struve to Francis Baily, Esq. President of this Society". The English is much better than Wilhelm usually wrote and Baily must have translated a German letter , or at least have edited an English one. The printed English text follows very closely Wilhelm's German account in Astronomische Nachrichten, Vol. 4, pp. 37-44 and 49-52, 1826, so it is used here (by permission of the Royal Astronomical Society) in lieu of a new translation.

7. H.C. King, 1955, The History of the Telescope, (ref. 2 above), p. 183, or see plate 2 in ref. 6 above.

8. J. von Fraunhofer, 1826, Astronomische Nachrichten, Vol. 4, pp. 35-38.

9. A. von Oettingen, 1894, Gedächtnissrede, p. 74.

10. J. von Fraunhofer, 1826, Astronomische Nachrichten, Vol. 4, pp. 17-28.

11. See ref. 6.

12. O. Pedersen. 1976, Vistas in Astronomy, Col. 6, pp. 17-28.

13. F.G.W. Struve, 1827, Catalogus Novus, p. II.

14. F.G.W. Struve, 1827, ibid, p. V.

15. F.G.W. Struve, 1827, ibid, p. V.

16. A. von Oettingen, 1894, Gedächtnissrede, p. 77.

17. L. Kellner, 1963, Alexander von Humboldt, Oxford University Press, pp. 132 and 144 give translated exerpts of these letters.

18. A.H. Batten et al., 1982, Publications of the Astronomical Society of the Pacific, Vol. 94, pp. 860-871.

19. O.W. Struve, 1895, Erinnerung, p.31.

20. O.W. Struve, 1895, ibid, pp. 32ff.

21. O.W. Struve, 1895, ibid, p. 33.

22. O.W. Struve, 1895, ibid, pp. 42-3.

23. O.W. Struve, 1895, ibid, p. 46.

24. O.W. Struve, 1895, ibid, p. 34.

25. J. South, 1833, letter of April 2 to F.G.W. Struve reprinted in Rara Astronomica in Estonia, Tartu Publications Vol. 55, pp. 44-5, 1977.

26. F.A. Mitchel, 1887, Ormsby MacKnight Mitchel: Astronomer and General, a biographical narrative by his son F.A. Mitchel, Riverside Press, Cambridge Mass. This extract has been reprinted in Pub. Yerkes Obs. Vol. 1, p. xiv, and in The History of the Royal Astronomical Society 1820 - 1920 p. 50 (edited by J.L.E. Dreyer and H.H. Turner, published by the R.A.S. London, 1923) which gives a fuller account of South's quarrel with Troughton and Simms.

27. W. Henop, 1931, Jacob Struve, p. 7.

28. O.W. Struve, 1895, Erinnerung, p. 48.

CHAPTER 6

THE FOUNDING OF PULKOVO

Naturally, Emilie's death marked a watershed in Wilhelm's life, but his life and that of his family would, in any case, have been radically changed by a chain of events set in motion some years before. Wilhelm himself tells us that after returning from Germany, France and England in 1830 (see Chapter 5):

> ...visiting the capital in the month of December, I had the honour to be admitted to the presence of His Majesty the Emperor, in order to give him an oral report on the results of my journey. On this occasion His Majesty was pleased to inform himself as to the condition of the Observatory of St Petersburg and I did not hesitate at all to give him a full account in complete frankness and according to the strictest truth.[1]

St Petersburg Observatory was, in fact, in a very bad state. It was part of a building that provided a home for the Academy of Sciences (to which it belonged), which had been built in 1725, at the end of the reign of Peter the Great. It was in the centre of the city and on top of the building's central tower, which could not properly support even the inadequate instruments available. The upper part of the tower was damaged by fire in 1747, but the Academy's astronomers still had to climb a long staircase to reach the observing floor --a circumstance that deterred one of the two elderly astronomers employed there at the end of the eighteenth century from making observations. No clearer testimonial needs to be offered to the inadequacy of the observatory than the dispensation Wilhelm received, on his election to the Academy, from the normal requirement that an Academician should live in St Petersburg --a dispensation grounded on the access he already had to the far superior instruments in Dorpat. The Academicians were well aware of the situation and planned to remedy it --but they lacked both money and a leader. Several instruments, including one of Herschel's twenty-foot telescopes were acquired in 1793, and three years later King George III of England gave Catherine the Great a 10-foot Herschel telescope. Catherine's plans to provide a new observatory were frustrated by her death. The great Swiss mathematician, Leonhard Euler, whom Catherine had persuaded to come to Russia, brought with him his compatriot Nicholas Fuss from Basel. Fuss married Euler's granddaughter, and edited the mathematician's works for posthumous publication; he, too, hoped to restore the observatory. Like Wilhelm, Fuss founded a scientific dynasty: his elder son Paul was permanent secretary of the St Petersburg Academy and Wilhelm's close friend. Paul's younger brother, Georg was Wilhelm's assistant both in Dorpat and in Pulkovo, and Georg's son Victor was a supernumerary astronomer in Pulkovo during the 1860s. Paul Fuss helped to realize his father's hope of restoring the Academy's observatory. By the time Paul and Wilhelm were active, it was obvious that a completely new building was needed, if the money to be spent on new instruments was not to

66

be wasted. The Academy negotiated with Count Kouchelev-Bezborodko, who had offered land on the north shore of the Gulf of Finland. The Tsar himself, however, had other plans.

The nature of Wilhelm's personal relationship with Nicholas I is worthy of a short digression because there is some evidence that it was important in the founding of Pulkovo. Although Wilhelm's grandson Peter played an important role in the early development of Russian Marxism, Wilhelm himself seemed well content with a system that had allowed him, the grandson of a barely literate peasant, to rise through the state service to the rank of an hereditary noble of the Russian empire. Otto tells us that he regarded the three monarchs whom he served --Alexander I, Nicholas I and Alexander II-- with "an enthusiastic respect".[2] Wilhelm's work in Dorpat had brought him favourably to the attention of both Alexander and Nicholas; his geodetic work, in particular, contributed to the growing respect with which the rest of Europe was coming to regard Russia. Wilhelm first met Nicholas in Dorpat when the latter --according to Otto, who gives no date-- came to see the Great Refractor which had become "a kind of wonder of the world which had to be gazed at by all educated men".[3] Georg Parrot, a confidant of Alexander and who knew Nicholas, probably also advanced Wilhelm's interests with successive Tsars. Nicholas is often thought of as one of the harsher members of a little-loved dynasty, but one of his recent biographers has pointed out that our opinion is largely formed from the writings of those exiled for their opposition to him.[4] At least in the early part of his reign, Nicholas was a competent ruler and ready to listen to those whom he judged loyal. Many of the "mid-level" managers in his civil service were, like Wilhelm, men of humble origin and from immigrant families. According to Queen Victoria, Nicholas' most obvious characteristic was a strong sense of duty.[5] He was likely, in his turn, to respect a man who had been brought up on the principles "each hour has its allotted task" and "work is the best and most useful spice of human life". Indeed, Alexander von Humboldt met Nicholas at the Prussian Court, while Pulkovo was still being built, and wrote to Schumacher:

> The Tsar has often and always with the most friendly interest spoken to me of his Observatory (the astronomical city whose form of government may be difficult to conceive) and of the worthy Struve.[6]

After listening, in 1830, to Wilhelm's frank description of the state of the St Petersburg Observatory, Nicholas declared that

> ...the honour of the country appeared to Him to require the foundation, close to the capital, of a new astronomical Observatory, suitable to the current status of the science and able to contribute to its further advancement.[7]

Prince Lieven, formerly the Chancellor of Dorpat University and then the Minister of Public Instruction was also at the audience and was told to busy himself with the matter "without delay". The Tsar even began to propose specific sites and, when Wilhelm told him of the offer of land from Count Kouchelev-Bezborodko

...the Emperor himself noted all the disadvantages, such as the proximity of a large city to
the south of the Observatory, and the poor security for the foundations that would be
afforded by the sandy and marshy soil on that side of the capital. His Majesty indicated,
at last, as appearing most suitable for this purpose the hill of Pulkovo, situated to the
south of the city and at a more considerable distance from its limits.[8]

In addition to Wilhelm's formal account of his audience with the Tsar, recorded in the *Description de l'Observatoire central de Poulkova*, and published in 1845, we have a rather more circumstantial one from Piazzi Smyth.[9]   In using the latter, allowance must be made for Smyth's affectation, already mentioned, of recording long monologues or conversations as if his account were *verbatim*.   This particular conversation took place after Wilhelm's serious illness that affected his memory (although not too badly for the distant past) and after a dinner at which wine was liberally served, but Smyth's account does not contradict Wilhelm's own and yet provides more detail.   According to Smyth, Wilhelm first asked for more money for Dorpat Observatory, and Nicholas countered by asking if it would not be better to build a large observatory near St Petersburg.   Wilhelm then explained the Academy's plans and Nicholas, raising his objections to the site offered by Count Kouchelev-Bezborodko, sent Wilhelm to inspect Pulkovo for himself.   When, at a subsequent audience, Wilhelm reported that Pulkovo was acceptable, Nicholas then sent him to look at several other sites, telling Wilhelm not to accept Pulkovo just because the Tsar had recommended it.   After looking at all these other sites, however, Wilhelm was convinced that Pulkovo would be the best. Cleveland Abbe[10] added that when Wilhelm first saw Pulkovo hill in 1828, he turned to his companion, Baron Wilhelm Wrangel, and exclaimed that upon that hill the new St Petersburg observatory would one day be built.

At first, no-one intended to buy completely new equipment for the observatory.   The 20-foot Herschel telescope, a 4-inch achromatic refractor and a recently bought Ertel meridian circle, all belonging to the Academy's old observatory were to be re-housed at Pulkovo at a cost of between 300,000 and 350,000 assignat roubles.   Other major instruments --including a 10.5 (Paris)-inch achromatic refractor-- were to be bought at a cost of around 120,000 assignat roubles.   The new Minister of Public Instruction, S. Uvarov, presented these plans to the Tsar who --on the 28th October 1833- approved the ordering of the instruments and released funds for the building.

This was the situation when Wilhelm felt obliged to leave his dying son in the care of his pregnant wife in order to attend a meeting in St Petersburg about the new observatory.   Uvarov --whom, like Nicholas himself, we usually see through the eyes of those who opposed him politically-- appears to have been genuinely committed to this project.   On October 31st he set up a commission of four academicians: de Wisniewsky, who had made several cartographic expeditions in the Russian empire, Paul von Fuss, Georg Parrot (later replaced by Lenz) and Wilhelm himself.   (Wilhelm does not specify whether this and other dates given in the next few pages are Old Style or New. Since he was writing in an official Russian publication, they are probably Old.) The Commission's chairman was Admiral Greig, a member of the Imperial Council, an honorary member of the Academy, who had made many surveying voyages and had built Nicolayev Observatory.   This Commission submitted detailed plans to the Tsar, who selected two architects --Bruloff and Thon-- to submit plans for the observatory building. Thon's plan, a Gothic facade, appealed to many members of the Commission, but Wilhelm persuaded them that Bruloff's design was more practical (although they insisted on some changes). At the same time, the Commission began to have much more ambitious ideas.

Wilhelm and Uvarov were received again by Nicholas on 15/3 April 1834 (Wilhelm's forty-first birthday) and explained to the Tsar the Commission's reasons for its choice of plans, and the new ideas. Nicholas designated Bruloff as the architect and Wilhelm as the Director of the Central Observatory, as soon as it should be built. Nicholas also commissioned Wilhelm to travel abroad to find the best possible craftsmen to make the new instruments. Otto says expressly that the Tsar used the phrase *carte blanche* in speaking of the cost of instruments,[11] and Wilhelm wrote:

> As to the quality of the new instruments to be ordered, His Majesty was pleased to make manifest to me his express wish, to see the new Observatory furnished with everything of this kind that was most perfect.[12]

In the meantime, Georg Fuss made the first observations on Pulkovo hill, beginning on the night of 28/16 March, to determine the position of the new observatory relative to the old, and so to establish Pulkovo's precise geographical position.

Thus, in June 1834, Wilhelm had to visit Germany and Austria in order to purchase instruments. This presented difficulties to a recently widowed family man. He had seven or eight living children to provide for (two-year-old Gustav died sometime in 1834). The eldest daughter, Charlotte, entered her Aunt Conradine's Educational Institute as a boarder. Alexandra and Olga (four and five years old) were looked after by the family of Wilhelm's friend and colleague Martin Bartels. (We are not told who looked after the infant Emilie, but she lived to a ripe old age --d.1912.) Three boys, Conrad, Heinrich and Bernhard were left to fend for themselves --possibly with some supervision from the older cousin Adolf-- while Otto accompanied his father. Otto had just turned 15; this is the first occasion that he records in the *Erinnerung* on which his father singled him out for any special attention. He had just completed his gymnasium course, and was to attend his father's classes during the next year, although he was not allowed to enroll officially in the University until after his sixteenth birthday. Otto's abilities were beginning to show, although, tantalizingly, he gives us no indication in material so far published how his own interest in astronomy began. He was now the oldest surviving son --which was apparently considered of some importance in the family-- but it is quite unclear whether his interest in astronomy was spontaneous or was deliberately fostered by his father in a conscious attempt to create and astronomical "dynasty". Wilhelm no doubt had practical reasons for taking Otto with him -- he would have been glad of a companion, and each of the children had to be cared for somehow. Whatever Wilhelm's motives, he introduced Otto to such scientists as Bessel, Encke and von Humboldt-- and whether by design or chance, the journey must have influenced Otto's choice of a career.

Piazzi Smyth wrote that Wilhelm was determined to tell the instrument makers exactly what he needed, and would not take just what they had available --as he had done with Fraunhofer's refractor. Wilhelm wanted both optical and mechanical perfection; not even a "single screw" was to be made without his approval. His most important port of call was Munich, where he visited Ertel -the successor of Reichenbach- and Merz and Mahler, successors of Utzschneider's optical institute where Fraunhofer had worked. The large achromatic refractor for Pulkovo, originally to have been 10.5 Paris inches in aperture, was now to be as large as possible. Merz and Mahler had just completed a 10.5-inch for an observatory near Munich and they invited Wilhelm to examine and use it. He did so under conditions as nearly as possible like those in Dorpat and concluded that the new instrument "was worthy of the successor of Fraunhofer, and that it united in a high degree the two essential qualities: precision of images and achromatism"[13] --although he made several suggestions for improving the mounting. Wilhelm encouraged Merz and Mahler to make an even larger objective, and they agreed to try to produce one of 13.5-inches aperture. Later, they thought they could do even better and eventually made an objective with an aperture of 14 Paris inches (or 14.95 English inches).

Wilhelm also visited the Repsold brothers in Hamburg, conveniently close to the home of his parents and parents-in-law. The firm of Repsold was founded by Johann Georg Repsold who had died a few years earlier. Two sons, and later two grandsons, continued the family tradition and, at different times, the firm traded as Repsold, Repsold and sons, or Repsold brothers. There was a family connection with this firm, since Kasimir Henop, Wilhelm's nephew, was a draughtsman with it. Later, Otto worked there for a period. Wilhelm had written to the Repsolds and they had detailed drawings --half natural size and showing every screw-- waiting for him. Wilhelm returned to St Petersburg having ordered most of the major instruments from Repsolds, Ertel, Merz and Mahler, Plössl (Vienna) and from many lesser known craftsmen. He could not, however, decide between designs for the vertical circle submitted by Ertel and the Repsolds, and presented both to the Commission. It decided unanimously in favour of the Repsolds' design because it included a new "admirable apparatus for reversing" it.[14]

During his travels, however, Wilhelm not only ordered instruments, he also sought the opinions of other astronomers on his plans for the new observatory. He showed these plans to Bessel, Encke, von Humboldt, Schumacher, Olbers, Lindenau, Steinheil and others. All, apparently, reacted favourably and Wilhelm came home much encouraged. He stopped in Königsberg on both the homeward and outward journeys and Besel's advice, no doubt, was particularly heeded.

At home, Wilhelm found the domestic situation far from satisfactory. The girls had been well looked after in his absence, but the boys --aged from 7 to 13 and left to themselves-- had quarrelled and encountered all sorts of difficulties. Perhaps Wilhelm, like many other brilliant scientists, did not always make the wisest decisions in everyday matters! He decided now, however, to contract a new marriage --to Johanna Bartels, fourteen years his junior, who is said to have been recommended to him by his first wife on her deathbed.[15] Johanna was the daughter of Wilhelm's colleague Martin Bartels,

Portrait of Martin Bartels (Reproduced by Permission of the Library of Tartu State University).

professor of mathematics at Dorpat since 1820. A document in the possession of the Struve family states that the French mathematician and astronomer, Pierre Simon, Marquis de Laplace, once said that Bartels was the greatest mathematician of the time, after Gauss. The story is repeated by Dunnington, in his biography of Gauss,[16] who also points out that it is told too of J.F. Pfaff (one of Bartels' teachers and elder brother of the Dorpat professor J.W.A. Pfaff mentioned in Chapter 2). Laplace was not given to modesty and might well have claimed the title for himself; if he did make the remark, there must have been good grounds for it. Although books written by Bartels are still extant, his name is not much remembered by modern mathematicians, but he is known to have taught both Gauss and Lobachewsky --a pioneer in the study of non-Euclidean geometry-- which, indeed, suggests that he had considerable abilities as a mathematician and teacher. Bartels was assistant to the teacher at one of the elementary schools in Brunswick where Gauss was a pupil. After Gauss' mathematical talent became evident, Bartels helped him and brought him to the attention of the Duke of Brunswick, who became his patron Lobachewsky was Bartels' student at the University of Kazan. Bartels was a native of Brunswick and his family can be traced there back into the late seventeenth century. In 1803 he married Anna Saluz from Chur in Switzerland. In five generations of her ancestors there were nineteen pastors.[17]   This element in Johanna's background seems to have been one of the strongest influences on her. She had a perhaps rather severe piety, but was nevertheless well liked by a wide social circle in Dorpat.   The young American astronomer Cleveland Abbe, who worked in Pulkovo for two years and knew Johanna near the end of her life wrote of her to his parents:

> I wish you could know Madame Struve. She is a quiet little woman but so good that everyone loves her.[18]

Wilhelm and Johanna were married on 22/10 February, 1835.   There was no strain between the children and their new stepmother for Otto writes

> ...he [Wilhelm] surely could not have given the children a better mother. You, my dear brothers and sisters, all in maturer years have known your honoured mother in her diligent calm, discreet and unpretentious activities, in her unequalled gladness of heart, and will, therefore, certainly consent to the above statement from full hearts.[19]

These words are further confirmed by a grandson of Olga Lindhagen who says that Johanna "was much loved by my grandmother, who was only three years old when her real mother died."[20]   Johanna soon had children of her own to care for too. The first, Karl, named after his uncle but often known in adult life by the Russianized name of Kyrill, was born in the November after the wedding. Friedrich and Paul, born in 1836 and 1838 respectively, each died in infancy. Johanna's father also died in 1838, but her only daughter, Anna, was born in the following year, just after the move to Pulkovo. Two other sons were born there, Ernst (1841) and Nikolai (1842).

On 15/3 July, 1835, the foundation stone of the new observatory was laid, after the singing of a *Te Deum* --a ceremony unaccountably omitted at the dedication of most modern observatories. This stone, laid in what were to become the observatory's subterranean cellars, formed part of the pier that would support the 15-inch refractor (as it is usually called). Wilhelm and the other members of the Commission were present, as well Uvarov and various state dignitaries. In the same year, Uno Pohrt, mechanic at Dorpat Observatory and destined to become head of the machine shop at Pulkovo, was sent to Munich to become familiar with the major instruments, to report on the progress of their construction and finally to accompany them on their journey. In 1836, work began on the three towers that would eventually carry the moving domes (or, rather, turrets). These turrets were to be similar to the one at Dorpat, except that they would ride on wheels independent of both the walls and the moving roof. The wheels were supported in a frame of their own between railways fixed to each of the other parts of the building. Before the lower rail was fixed to the walls, Wilhelm wished to be sure that their top was level. Various fixed points on the walls were measured throughout the changing seasons of two years. The north wall of each tower did subside relative to the south but in no case by more than six hundredths of an inch. The westernmost turret was put in position first, on 19/7 June 1838.

In August 1838, word came from Munich that the instruments were complete; Wilhelm was invited to come inspect them before they were shipped.

Once again, Otto accompanied his father. This journey proved to be Wilhelm's last opportunity to see his own father (d.1841) and Bessel (d.1846); his brother Karl had died earlier that same summer. In Munich, Wilhelm was joined by Encke from Berlin and both of them (and Otto as well) made trial measurements with the Reichenbach (or Ertel) meridian circle. Their vernier readings all agreed within 0".16, but Wilhelm found that if he used a microscope, his own readings differed by as much as 1" unless he took care that the scale was always illuminated from the same direction. Merz and Mahler had been required by the contract to build temporary housing for the two major telescopes so that they could be tested under conditions similar to those in which they would be used. After five weeks in Munich, Wilhelm found the instruments satisfactory, except for some minor changes. He then visited Repsolds' in Hamburg and Voss, the Academy's bookseller in Leipzig, with whom he had arranged the purchase of many volumes for Pulkovo's library.

During all this activity, Wilhelm was still teaching at Dorpat and using the Fraunhofer refractor there. It was almost impossible to understand how he found time for his duties as a university professor while meeting these new responsibilities of supervising the construction of Pulkovo. He had to make many journeys to St Petersburg at a time when railways, still a novelty in England --their land of origin-- were hardly known at all in Russia. Roads to the capital were poor and the journeys time-consuming. In addition, just at this time, Wilhelm was instructing G. Fuss and Sawitsch in the methods to be used in the levelling surveys between the Black and Caspian Seas. At last, after 31 years in Dorpat, the time came for him to leave. His appointment as Director of the Central Observatory came into effect on the 13/1 April 1839, and six days later he arrived to take up residence there. In the fall of 1838, the University Council had made him Professor Emeritus, thus recognizing 25 years of service, but asked him to continue, for the time being, his accustomed duties. Now Wilhelm asked to be allowed to leave before the end of the semester, just as, 31 years earlier, he had left the Christianeum without completing the course. He took his leave of his *Alma Mater* in a letter dated 8 March, 1839:

> To be sure, I would still like to stay in my present activity until the end of the current semester , if the work beginning in April at the Central Observatory did not require my presence and immediate direction. In the meantime, it contributes to my peace of mind to be able to assure the esteemed Council about the interruption of my lectures in the middle of the semester, that by increasing the hours for some time until Easter, I have brought the more important of the two courses "On the Calculus of Probabilities" completely to an end, and also the other "On Spherical Astronomy" is in its principal parts completed.

> Honourable as is the new office to be entered upon, yet the imminent separation from Dorpat must arouse in me the saddest feelings; still more, however, feelings of gratitude to the University to which I have belonged, uninterruptedly, for nearly 31 years, as student and teacher. I came to Dorpat University in the year 1808, as a raw youth, and to it I owe not only my first scientific education, but also, after I had exchanged philological studies for mathematical ones, the means to continue my studies at the University. It was Parrot the father, in particular, who recognized and encouraged the far from usual zeal of the young astronomer. He was my benefactor, since he prevailed on the Council, without my asking --indeed only suspecting the grant, to award to me considerable support, so I could devote myself to the new subject completely at leisure. Two years later (1813) the Council honoured me with its consideration in filling the position of Extraordinary Professor and

Lithograph by G.F. Schlater from a portrait of Wilhelm Struve painted by Eduard Hau some time between 1836 and 1839 The quotation from Seneca (in Wilhelm's own hand) may be translated: There are vast spaces above, into the possession of which the mind is admitted From a copy provided by the Tõravere Observatory

Observer at the Observatory, thus opening to me the path of an academic teacher and making astronomy the mission of my life. More than 25 years have since flowed by. As I look back in my mind on this interval, how many reasons do I not have to be thankful next to God, to the authorities of the University, and especially to the Council, for so much good experienced by me personally and by the institute entrusted to my leadership, the Observatory? When I took up my office, the Observatory had just been built and furnished only with modest scientific apparatus. How often did I not come to the Council with the request to obtain some extension of this, thus gradually bringing the Observatory into the ranks of those completely equipped? Each request received attention; each, through the representation of Council, [received] a grant or patronage from the authorities. Means were provided for publishing the works of the Observatory, for scientific journeys, and undertakings of the larger kind, such as the degree measurement, were carried out, so now Dorpat University stands richly equipped, as few in the world, well-known and highly esteemed.

It is the spirit of forwarding at all costs the scientific interests, on the part of the Council and their superiors,* that has made the history of Dorpat University glorious! May this spirit characterize Dorpat University for all time to come.

Not only in scientific respects, however, but also in purely human ones, I have richly experienced good in Dorpat. Accept, honoured, beloved, older and younger colleagues, the thanks of the friendly-disposed man who is leaving you --for the many proofs of trust and friendly benevolent feelings. Believe the assurance that their memory stays inextinguishable in the heart, and grant this sincere request that I, the distant one, may still be seen to belong to your circle.

Professor Emeritus W. Struve[21]

Although the words may seem high-flown to our modern ears, they were not merely the expression of nineteenth century sentimentality. Wilhelm made similar remarks in French in the *Description de l'Observatoire Central de Poulkova* and leaving Dorpat was obviously a major turning-point in his life. The greater part of the credit for the reputation of Dorpat Observatory belongs to Wilhelm, despite his generous acknowledgements of support. He did indeed leave behind many friends amongst his colleagues: Otto wrote that "It does not belong to my recollection that serious ill-feelings lasted for long between him and any one of his Dorpat colleagues".[22] While there might have been quarrels when Otto was too young to remember them, Wilhelm's friendship with Bartels and both Parrots, as well as with many others, are evident from the present account. He seems to have been one of the "characters" of Dorpat University and figures in a collection of accounts of such people published shortly after his death:

Struve was a powerful, almost Goethean figure. He fitted exceptionally well with his very beautiful first wife who manifested the noblest oriental type. His second wife was the daughter of the famous professor of mathematics, Bartels, who once, at the school at Altdorf, had Louis Philippe as successor. On this account, Professor Erdmann said at Bartel's funeral: "It is hard to say if this circumstance redounds more to the honour of Bartels or of the King or of the French".[23]

---

* Emphasis found in the source.

This reminiscence continues with an account, designed to illustrate Wilhelm's ready wit, of an encounter between him and the professor of philology.

The Dorpat years had to come to an end, however, and the University's loss was at least as great as Wilhelm's. Three of his assistants (and former students), namely G. Fuss, Sabler and Otto, were appointed to the positions of associate astronomer at the new observatory. Pohrt the machinist was taken, at first in a temporary capacity, as was another of Wilhelm's students, Döllen, who later became his son-in-law. A temporary appointment was also given to Lt. Christian Andreas Schumacher of the Danish Navy. This nephew of H.C. Schumacher later worked as Alexander von Humboldt's assistant. Dorpat, astronomically speaking, was quite emptied by Wilhelm's move, and in April the observing assistant, Preuss, died. The quality of the Observatory's instruments ensured that it remained important throughout the nineteenth century, but the greatness of Dorpat University departed with Wilhelm. Only in recent times has it recovered some of that quality with the foundation in 1964 of a new observatory, named after Wilhelm, at Tõravere -some 20 km from Dorpat, or, as it is now called, Tartu.

Wilhelm's last recorded observation from Dorpat was of occultations of stars in the Pleiades on the night of 19 March 1839, and was made with a group of five colleagues. It was typical of Wilhelm's interests since, if similar observations were made elsewhere, the results would enable the differences of longitude between observatories to be worked out. This was one way, before the use of the telegraph became general, of refining determinations of longitude. Group observations of this kind were common in Dorpat --more than one visiting astronomer found himself pressed into helping-- and were continued at Pulkovo. Piazzi Smyth records taking part in such an effort in the latter place, as late as 1859. The Pleiades were particularly useful for this purpose because, as the Moon passes between us and the cluster, it usually occults several of the stars in quick succession, so many individual timings can be made --thus improving the longitude determination. Because of the characteristics of the Moon's motion, occultations of the Pleiades occur in cycles, a number being observed in two or three consecutive years, and then no more occurring for nearly nineteen years. Piazzi Smyth's observations were also of stars in the Pleiades --just one cycle later than Wilhelm's last Dorpat observation.

The tasks requiring Wilhelm's immediate attention at Pulkovo were the erection of the two remaining domes and the installation of the instruments. He arrived early in April and was expected to have the Observatory ready for formal opening by the end of July. The domes were completed in May, but most of the instruments still had not arrived at the beginning of June. Those from Munich were sent in three wagons, one of which (carrying the lenses, divided circles and other sensitive parts) was especially sprung. Under Pohrt's personal supervision they were taken from Munich to Lubeck without change of horses or drivers. At Travemunde, these instruments were joined by clocks and other accessories made in England, and all loaded on a steamboat for St Petersburg. At the mouth of the Neva they were transferred to barges and

Pulkovo Observatory from the north (nineteenth-century photograph).

taken through the city by canal to the highway, arriving at Pulkovo on 8 July/26 June. The instruments made in Hamburg came by a similar route somewhat earlier. At last, except for a few still being made, the instruments lay on the floor of the central room of the Observatory, packed in 102 crates. "To our great satisfaction" wrote Wilhelm "the opening let us see that all the instruments had arrived in a state of repair that left nothing to be desired; that is to say without the least damage to any part".[24]

The manufacturers were to have sent people to help to install the instruments, but Wilhelm working against a deadline --or perhaps just impatient as he always was when he had a new instrument to play with-- decided to proceed at once. Within six weeks, the associate astronomers, Pohrt and his assistant Wetzer --all working under Wilhelm's "immediate direction"- had assembled all the instruments in their final positions. One would expect no less from the youth who had ground holes in the granite pillars that were to support the Dollond transit, or the young professor who had assembled the Great Refractor with no guidance but "a perspective view...which represents the view from the side on which the clock is situated".

The date for the formal opening, August 19/7, 1839, could now be set By order of the Tsar, all Russian astronomers were summoned to take part: they came from Moscow, Kazan, Kiev, Kharkov, Nicolayev, Vilna, Dorpat, Mitau (Wilhelm's oldest friend Paucker), Helsingfors, (Helsinki) as well as from St

Petersburg itself. Uvarov attended as Minister of Public Instruction and as President of the Academy --which body came in full strength. High officials, foreign ambassadors and distinguished scholars were invited. Admiral Greig and the Observatory Commission, of course, were there: "a brilliant company assembled at Pulkovo towards 11 a.m."[25] The Admiral formally handed the Observatory over to the Minister. Divine Office was celebrated in the central room and the entire building was solemnly consecrated. The Director then made a fitting speech, explaining the scientific importance of the occasion and expressing gratitude to the august Founder, the Minister, the President of the Commission and the architect. Then he reminded the astronomers of the Observatory of the serious duties imposed upon them by "the interests of science and the honour of the fatherland"[26] and invited all astronomers of the Empire to work together with the Central Observatory "in a manner worthy of the encouragements --almost without precedent-- that the government had accorded to this sublime science".[27] Copies of a medal specially struck for the occasion were distributed to persons selected by the Minister. The medal bore the effigy of Nicholas and the inscription (in Russian) "D.G. Nicholas I, Emperor and Autocrat of all the Russias" and on the reverse a representation of the building, surrounded by the signs of the zodiac and with the inscription "The year 1839. By order of Emperor Nicholas I". The guests then dispersed through the various rooms of the Observatory and finally were offered refreshments --the weather being fine-- in a tent outside the front door.

A few days later there was more excitement when the Empress, accompanied by the Grand Duchess Olga, "was pleased without preliminary announcement, to honour the Observatory with her visit" (one imagines that this was not quite the phrase that leapt to Wilhelm's mind as he saw the Imperial carriage roll up outside the Observatory door) "and to inspect the large telescope and the mechanism of the big dome".[28] Nicholas' own visit was later (8 October/26 September) and carefully planned. He arrived with a considerable and distinguished entourage, including Lt. General Schubert (commander of the marine longitude survey in which Wilhelm had participated in that fateful summer of 1833). The visitor spent several hours in a grand tour of the entire establishment, from the basement to the central dome with its 15-inch refractor.

> At his departure, the Emperor was pleased to express His great satisfaction in the most flattering terms; His Majesty congratulated Mr. Uvarov that this great foundation had been conceived and built during his ministry; He was pleased to address to the Director the question of whether or no there still remained something that he desired. Encouraged by this new mark of august solicitude, the Director protested that, thanks to the munificence of His Majesty, he could not have any new wish, unless it be to see His Imperial Majesty keep for ever his great favour toward the new-born establishment and toward those called to work there. He implored His Majesty to be pleased to permit, if in the future some new need should come to be felt, that then the Observatory might dare to have recourse to the generosity of its magnanimous founder.[29]

The Tsar did grant the Observatory the right of appeal to himself --a privilege prized as greatly as it was little used.

Obverse and reverse of medals struck on the occasion of the dedication of Pulkovo Observatory. (See text for the inscription.)

The village of Pulkovo, situated just off the main highway to Moscow, about 20 versts from St Petersburg, contained some 2,000 souls and was part of the Imperial estate of Tsarskoe Selo. The village itself is at the bottom of the hill on which the Observatory stands. The ground level at the entrance staircase of the Observatory is --according to a levelling survey by Otto Struve and Liapounov conducted in 1834-- 242.8 feet above the mean level of the Gulf of Finland, 14 versts away. St Petersburg was built on a marshy plain and there are not many hills of even this elevation nearby; the astronomers enjoyed a good view of the city, almost every public building of the time being visible The American astronomer Simon Newcomb relates that Otto told of being able to watch (through the 15-inch refractor) sailors on the decks of British ships blockading Kronstadt (25 miles or 40 kilometres away) during the Crimean War.[30] The Observatory is still visible from the highway, once one has left the city, and this commanding position cost Pulkovo dear in a later and much grimmer war, when St Petersburg had become Leningrad and was subjected to a long and terrible siege. In the happier times of its earlier years, however, Pulkovo found it an advantage to be raised above the smoke of the city's chimneys and the fog that so often hung over the marshland.

The Observatory was a magnificently conceived building, every part designed to serve the central purpose of astronomical observation. The cellars or "subterraneans" were just as important as the superstructure. Their principal purpose was to protect from temperature changes the great stone masses, extending 30 feet into the ground, on which the piers of the various instruments were to rest. Summer temperatures could be as high as 88°F (31°C) and winter ones as low as -24°F (-31°C). Each of these masses was surrounded by double walls going down almost as deep as the stone blocks themselves. The cellars also housed the central-heating stoves and served as a storehouse for

The 15-inch refractor shown on the new Repsold mounting installed in 1880. (Photograph from a collection distributed by Pulkovo about 1930 and in the library of the Dominion Astrophysical Observatory.)

The main building of Pulkovo from the south. The large screen on the projecting portion of the building protects "Wilhelm's" prime vertical instrument from direct sunlight. (Source as for the photograph on p. 80.)

food --very necessary for a community likely to be isolated during the Russian winter. These cellars were under the main part of the building --the "Observatory properly speaking" as Wilhelm called it. Above ground, this was a two-storey building crowned by the three domes (or turrets) housing the equatorially mounted instruments. The big central turret, of course, housed the 15-inch refractor. In the eastern turret was a $7\frac{1}{2}$ -inch heliometer, modelled on the instrument with which Bessel had successfully first measured the annual parallax of a star, but which, by Wilhelm's own account in 1845, had not been much used. The western turret had been intended for a refractor of six or seven inches aperture that Plössl of Vienna had contracted to make but was too old and ill to complete. Later the turret housed other instruments.

The two side turrets were offset with respect to the central one, to minimize the extent to which any one of them would block the view of the other two. The building joining them, however, lay east-west and contained observing rooms for the transit instruments. These rooms were carefully isolated from the central-heating system, but could be opened up on ceremonial occasions --such as Nicholas' visit-- so that distinguished visitors could walk the full 230-foot length of the building. Another wing, 172 feet long, extended southwards from the main tower and contained the observing room for the prime-vertical instrument and the library (which could also be extended upstairs). The astronomers lived in separate apartments --a two-storey one being reserved for the Director-- which were joined to the main building by closed walkways that could be heated in winter. There were also separate buildings in which small portable geodetic instruments could be used and tested.

Under the main tower there was a reception room for ceremonial occasions, which also did duty as a store-room for portable instruments. The Observatory's master clock was originally housed there, let into a niche in a

The Repsold transit instrument   (Source as for the photograph on p 80)

Wilhelm's prime vertical instrument by Repsold. (Source as for the photograph on p. 80.)

Jensen's portrait of Wilhelm Struve painted in 1841-2. Copy provided by Pulkovo Observatory.

granite pillar, as isolated as possible from temperature changes. Later the clock was moved to the more thermally stable basement. In another niche, facing the main entrance, was a bust of the "August Founder" Nicholas I. Around the walls of the room were portraits of famous astronomers, living and dead. All the portraits were either originals, or copies of the best originals extant. Many were commissioned from and painted by the Danish portraitist C.A. Jensen, who appears to have had a special gift for portraying scientists. He painted Airy, Bessel, Gauss, Hansteen, Sir John Herschel, the Repsold brothers, Schumacher and South, and copied portraits of Tycho Brahe, Newton, Flamsteed, Halley, Bradley and Roemer. Whether the idea for the gallery was Wilhelm's or Jensen's (as the latter's biographer suggests) is unclear, but Wilhelm's own portrait was the centrepiece. Jensen came to Pulkovo twice, and painted two portraits of Wilhelm Struve. He made a copy of the first one for the King of Denmark (Wilhelm was born a subject of the Danish King) which, although it is not considered as good as the original, may still be seen on request at Frederiksborg Castle in Denmark. Many modern visitors to Pulkovo are still impressed by the unique collection of astronomical portraits. Jensen also copied some of his other portraits for other institutions.

The library is one of Pulkovo's crowning glories. Wilhelm went to much trouble to assemble a complete collection, and with Nicholas' *carte blanche* behind him, he achieved a large measure of success. He had first choice of the books on mathematics in the library of his late father-in-law, Martin Bartels, and C.A.F. Peters of Hamburg --the fourth of the associate astronomers- sold some of his personal library to the Observatory. The Academy in St Petersburg gave a complete set of its memoirs and its collection of Kepler's manuscripts; the Prussian Academy in Berlin gave a set of *its* mathematical and physical memoirs. Kepler's manuscripts had been purchased by Catherine the Great on the advice of Euler. Later, Otto as Director was to afford opportunities for their study and editing, and, indeed, contributed to the latter himself. Fortunately, before the Second World War, the manuscripts were returned to the Academy and to Leningrad (St Petersburg). Some account of their history is given by Mikhailov[31] and Raskin[32].

Wilhelm began to assemble a collection of books for the Central Observatory's library as soon as it had been decided to create such an institution; in particular --as we have seen in Chapter 5-- he gave a complete set of Herschel's papers, annotated in that author's own hand. The Observatory at Vilna ceded a number of classical works, some very valuable, to the Central Observatory and many foreign institutions and academies made gifts. The most important single collection acquired, however, was the personal library of the German astronomer Olbers. Wilhelm had known Olbers personally, and in 1841 was given the opportunity to buy the latter's library after his death. So magnificent was the collection thus built up that Wilhelm wrote:

> ...I flatter myself, [it] can already be ranked amongst the most complete of observatory libraries.[33]

Of course the library continued to grow, acquiring many observatory publications by exchange and buying new books as they appeared. Fortunately

The Kepler manuscripts. (Source as for the photograph on p. 80.)

much of this great wealth survived the destruction of the Observatory in 1941, since the historically important material had been moved to Leningrad for safety. Pulkovo Observatory was, and still is, a good place to study the history of astronomy.

What did it all cost? Wilhelm estimated the total cost of building, instruments, books, etc. at 600,000 silver roubles.[34] This figure did not include the value of the land made over to the Observatory by the Tsar.

NOTES

1.   F.G.W. Struve, 1845, Description, pp. 24-25.

2.   O.W. Struve, 1895, Erinnerung, p. 55.

3.   O.W. Struve, 1895, ibid., p. 33.

4.   W. Bruce Lincoln, 1978, Nicholas I: Emperor and Autocrat of all the Russias, Indiana University Press, p.9.

5.   W. Bruce Lincoln, 1978, ibid., p. 196.

6.   K.-R. Biermann (ed.), 1979, Briefwechsel zwischen Alexander von Humboldt und Heinrich Christian Schumacher (see Chap. 3, ref. 9). Quoted with permission. The German original gives the last phrase as "und von dem wackeren Struve".

7.   F.G.W. Struve, 1845, Description, p. 25.

8. F.G.W. Struve, 1845, ibid., p. 25.

9. C.P. Smyth, 1862, Three Cities in Russia (see Chap. 4, ref. 8) Vol 2, Chap.3.

10. Cleveland Abbe, 1867, Dorpat and Pulkowa, Report of the Smithsonian Institute, Appendix, pp. 370-390 (p.373).

11. O.W. Struve, 1895, Erinnerung, p. 49.

12. F.G.W. Struve, 1845, Description, p. 29.

13. F.G.W. Struve, 1845, ibid., p. 34.

14. F.G.W. Struve, 1845, ibid., p. 34.

15. O.W. Struve, 1895, Erinnerung, p. 50.

16. G.W. Dunnington, 1955, Carl Friedrich Gauss, Titan of Science, (see Chap. 4, ref. 11) p. 255. Other details of Bartel's life may be found in E.T. Bell, 1937, Men of Mathematics, Simon and Schuster, New York, pp. 222-4 and in the Neue Deutsche Biographie.

17. O. Tomaschek, 1960, Ahnenlisten, p. 10.

18. Cleveland Abbe, 1865, letter of 20/8 October to his parents. Library of Congress, Cleveland Abbe Papers, container 2.

19. O.W. Struve, 1895, Erinnerung, pp. 50-51.

20. Nils Lindhagen, 1983, private communication.

21. A. von Oettingen, 1894, Gedächtnissrede, pp. 85-6, reproduced by permission of the Astronomische Gesellschaft.

22. O.W. Struve, 1895, Erinnerung, p. 46.

23. Dr. Bertram, 1868, Dorpat Grössen und Typen vor Vierzig Jahren, W. Gläsers Verlag, Dorpat, p. 28.

24. F.G.W. Struve, 1845, Description, p. 47.

25. F.G.W. Struve, 1845, ibid., p. 47.

26. F.G.W. Struve, 1845, ibid., p. 47.

27. F.G.W. Struve, 1845, ibid., p. 47.

28. F.G.W. Struve, 1845, ibid., p. 48.

29. F.G.W. Struve, 1845, ibid., p. 48.

30. Simon Newcomb 1903, Reminiscences of an Astronomer , Houghton Mifflin and Co., The Riverside Press, Cambridge Mass. p. 312. Newcomb made a similar remark in A Compendium of Spherical Astronomy, Macmillan and Company, 1906, p. 306.

31. A.A. Mikhailov, 1975, Vistas in Astronomy, Vol. 18, pp. 933-935.

32. N.M. Raskin, 1975, ibid., pp. 937-943. See also O.W. Struve, 1860, Mem. Acad. Impériale des Sciences, St. Pétersbourg, Sér. 7, Vol. 2, No. 4, p. 3.

33. F.G.W. Struve, 1845, Description, p. 257.

34. F.G.W. Struve, 1845, ibid., p. 53.

Heinrich Christian Schumacher   Lithographed by C Raugniet "from a picture in the possession of the Royal Society"   Reproduced with the permission of the Royal Astronomical Society

CHAPTER 7

THE ASTRONOMICAL CAPITAL OF THE WORLD

The move of the Struve family to Pulkovo marked Otto's scientific coming of age. Not quite 20 at the time, he was the youngest of the associate astronomers. In Dorpat he had been his father's student and --like Sabler and Fuss before him-- very lowly assistant. In Pulkovo he was entrusted with the brand-new 15-inch refractor, with which he looked for any double stars his father might have missed, followed unknown or suspected orbital motions of others, and surveyed the whole northern sky down to stars of the seventh magnitude. Already in 1845, Otto officially became Deputy Director and took over much of the internal administration from his father. When the time inevitably came to transfer full responsibility from father to son, there was little outward change. Otto remained Director until just after the Observatory's fiftieth anniversary, and the continuity in Pulkovo's life during those fifty years was not much affected by the formal change of Director in 1862.

Simon Newcomb appears to have been the first to quote the description of Pulkovo --ascribed to Benjamin Apthorp Gould-- as "the astronomical capital of the world".[1] Gould, who founded the *Astronomical Journal* (modelled on the *Astronomische Nachrichten*), studied in Europe under Wilhelm's friend Argelander and may well have coined the phrase. It was an apt enough description, but the capital was rather far removed from most of its provincial cities. Travel to St Petersburg from the cultural centres of western Europe was neither cheap nor easy, and was still more difficult for the just-emerging scholars of America. For example, no trains ran from Berlin to St Petersburg until after Wilhelm's death --yet visitors came, even in the early days. From their accounts (both in Wilhelm's day and Otto's) and from the Struves' own letters, we can form some idea of life in this "astronomical capital".

Fittingly enough, Wilhelm's old friend Schumacher was (in 1840) the first important foreign astronomer to make the journey. In the *Astronomische Nachrichten* he wrote of his impressions and gave accounts of each of the principal instruments, who was working with them and what they were doing. Until Wilhelm published fuller accounts in the *Description* in 1845, Schumacher's article contained all that most astronomers knew about Pulkovo He described Wilhelm's work on the constant of aberration with the prime-vertical instrument, Sabler's work with the meridian circle, Peters' with the transit and vertical circle, Fuss' survey of the northern sky with the heliometer and Otto's work already mentioned. Clearly impressed, Schumacher wrote

It is scarcely possible, without having been there, to have an adequate idea of the sublime beauty of the building and of the serious magnificence, worthy of science, of the interior furnishings, while any pointless luxury is disdained, nothing is spared for the safety and

89

convenience of observation. Still less can be described in words the spirit of order and
integrity that Staatsrath v. Struve [i.e. Wilhelm] has introduced into this great whole and
knew [how] to keep in it...

Over the portal of the main entrance one sees only the year of completion. No inscription -
-so the exalted Founder wished it-- announces His name. Frankly, where the work speaks,
no inscription is needed, and never will the gratitude of astronomers forget who erected
this admirable temple to their science.[2]

The next important visitor was George Biddell Airy who in 1836 had
been made Astronomer Royal and Director of Greenwich Observatory. He was
an outstanding scientist, who served Greenwich well --leaving that Observatory
much better equipped than he found it-- and whose work on the theory of
optics is of classical importance. The negative role he is widely believed to
have played in the discovery of Neptune in 1846, and certain of his personal
qualities, still injure his reputation. A hard worker himself, he expected his
assistants to work equally hard --but he did not trust them and even for the
simplest jobs, he wrote out detailed instructions, which he expected to be
followed to the letter. He was meticulous to a fault. Every letter he received
and a gelatine copy of every one he sent were filed and preserved. It has been
said that, if he wiped his pen, he would file the blotting paper![3] The contents
of the Airy archives show that at worst this is a pardonable exaggeration. His
own staff found him well-nigh impossible to work for, but those whom he
regarded as equals --mostly the directors of foreign observatories-- saw a
different side of him. They did not merely respect him for his work and the
position he held; they obviously regarded him as a true friend.

The Struves' friendship with Airy was particularly warm. Wilhelm first
met him in England in 1830 (Chapters 4 and 5), but the two men did not
become intimate until Airy, on Wilhelm's proposal, was elected a foreign
corresponding member of the St Petersburg Academy. Wilhelm's first surviving
letter to Airy is a semi-formal one in French, dated January 1841, in which he
informed Airy of the election. Wilhelm explained that he had nominated Airy
in recognition of his scientific achievements and also to create an opportunity
for them both to enter into regular correspondence. He also invited Airy to
Pulkovo, reminding him of Schumacher's visit and expressing the hope that
Airy would be the second foreign astronomer to travel there. Airy replied
appropriately in English. After one more effort in French, Wilhelm wrote on
September 14, 1841:

Dear Sir,

You will excuse, when I risk to write you an English letter, perhaps very badly composed;
but I find it more convenient to answer to an Englishman, when possible in his own
language, than in a tongue strange to both writers.[4]

Thereafter Wilhelm and Otto always wrote in English to Airy. Their
letters were not "very badly composed" but neither of them was completely at
home with English idiom or spelling --although their meanings are usually
clear. The friendship and correspondence lasted beyond Wilhelm's death, when
Otto continued it, until Airy himself died. Otto's eldest son, Georg Wilhelm,

Sir George Biddell Airy, seventh Astronomer Royal of England. (Reproduced with permission from the collection of presidential portraits of the Royal Astronomical Society.)

was Airy's godson. Despite this intimacy, Airy always began letters, either to Wilhelm or Otto, with "Dear Sir" or "My Dear Sir". Wilhelm once tried "My Dear Airy" but still got back the more formal salutation. Otto, who wrote to so many correspondents as "Theurster Freund" sometimes tried "My Dear Old Friend" on Airy (particularly when writing a letter of condolence on the death of Lady Airy), but he too always received the formal reply.

It took some years to persuade Airy to make the trip. In the letter quoted above, Wilhelm expressed regret that Airy had been unable to come that summer and renewed the invitation for 1842. Otto joined in the persuasion, but he was first to meet Airy in England in 1844, when Otto and Döllen determined the longtitude difference between Greenwich and Altona (and therefore Pulkovo --see Chapter 9). Early in 1847, Wilhelm was still trying to persuade Airy, recommending him to travel by Copenhagen, Stockholm and Helsingfors, in which last city he could

> ...see a very nice observatory and, a very intelligent astronomer M. Woldstedt, married to a niece of mine [a daughter of Wilhelm's brother Ernst]. There is likewise a very distinguished philosopher Mr. Nerwander, who works emminently well for magnetism and meteorology. I mention all what could help to call You, if Pulkowa alone is not sufficient. But I count most upon the help of Msstr. Airy, whose kind expressions annexed to one of Your letters have really moved me the heart.[5]

A few days later, Otto once again seconded his father:

> For the coming summer I will stop all time quietly at Pulkova, to finish and publish the calculations on the last expeditions. I must confess that the hope to see you here, has very much contributed to engage me not to leave Pulkova in the coming summer. It is this one of my happiest thoughts that perhaps we may have the pleasure to see in Pulkova the Astronomer Royal of England and his highly venerated lady. As you have already tried with good success the North Sea, I think our Baltic Sea cannot be more an object against the execution of your friendly intentions.

> Your little godson has passed these last four weeks in the bed. He was attacked, together with his sister, by the scarlet fever, but already the[y] are both in recovery. Our Georg William was the first two years of his live [sic] very often and dangerously ill from a scrophulous disposition of his body. Now at last it appears he has overcome the sickliness and this last half year, with his physical constitution also his intellectual capacities have considerably been developed. I hope therefore that, when you come here, you will find your godson a well constituted little fellow.[6]

In the summer of 1847, Wilhelm had to go to England to arrange the shipping of the standard measures for which he had asked in 1830. He saw his mother for the last time while on his way to England; she died before he was in Altona again on the return trip. More details of the visit to England are given in Chapter 9. Airy joined Wilhelm in Altona (without his "highly venerated lady") and at last accompanied his friend to Pulkovo. (Airy's own account differs in some details from that given by Otto in the *Erinnerung*). Airy also wrote his impressions (in English) for the *Astronomische Nachrichten*, from which the following extracts are taken:

> As the distance of Pulkowa from the principal scientific institutions of other countries in Europe causes it to be little visited by foreign astronomers (for I believe that you [Schumacher] and I are the only superintendents of foreign observatories who have travelled there) I have thought that an account of the Observatory, or at least of my general impressions of it, might have some interest for the readers of the ASTRONOMISCHE NACHRICHTEN;...

I will not longer delay the general expression of the extreme gratification which I have derived from my examination of the Observatory From the remarks which I subjoin I think that you will perceive that I have not been led on simply in blind admiration of every thing before I had examined it   I trust therefore that you will not regard as unfounded the opinion which I express, that no astronomer can feel himself perfectly acquainted with modern observing astronomy in its most highly cultivated form who has not well studied the Observatory of Pulkowa

To this excellence many antecedent circumstances have materially contributed The first of these is the personal character of Mr Struve, The third point is the almost perfect freedom as regarded the form of connexion with the Government, the selection of a department of astronomy to which the energies of the Observatory should be devoted, and the selection of a locality and of new instruments for the Observatory I am (as you know) connected with a more ancient establishment, and though I am far from complaining of all the necessary results of such a connexion, and though everything practicable is done to remove any inconveniences which may still seem to be attached to it, yet I cannot help feeling that faults in the selection of localities or in the plan of buildings which formerly were unimportant may now become serious, that it is difficult to abandon the use of an antiquated instrument and to substitute for it a modern construction, and that even in official and personal arrangements, ancient systems may occasionally be injurious to the efficiency of the institution From these historical inconveniences Mr Struve has been free, and he has admirably used his freedom 7

After giving some details of this admirable use of freedom, Airy reveals something of his own character:

the Observatory of Pulkowa is connected with the Astronomical Survey of the Russian empire, and here also is a certain analogy with the Observatory of Greenwich, which is connected in various ways with the Nautical Astronomy of Great Britain It is, in my opinion, exceedingly important, as well for preventing astronomers from wasting their time in the mere fanciful abstractions of science, as for giving to the Observatory its proper place in public estimation, that it should be in part devoted to some distinctly useful purpose of this kind 8

After some more similar remarks, Airy described the general impression that the buildings had made on him, echoing, to some extent, Schumacher's own thoughts:

The first thing which claims the attention of persons who have examined the building, as I have, is the care taken for the foundation of the instruments Little do persons who merely look in a cursory way at the Observatory, or who have not studied the sections in Mr Struve's description of the Observatory suspect what masses of brickwork are concealed beneath their feet

The next striking point is the general aspect of the structure of the Observatory Of this I may say generally, that while it is well-constructed, --even nobly and expensively constructed-- it is not constructed extravagantly, or with a disregard of expence while on the one hand it is worthy of an Emperor, it is on the other hand built in a style which ought not to be displeasing to a Minister of Finance 9 [Punctuation as in the original]

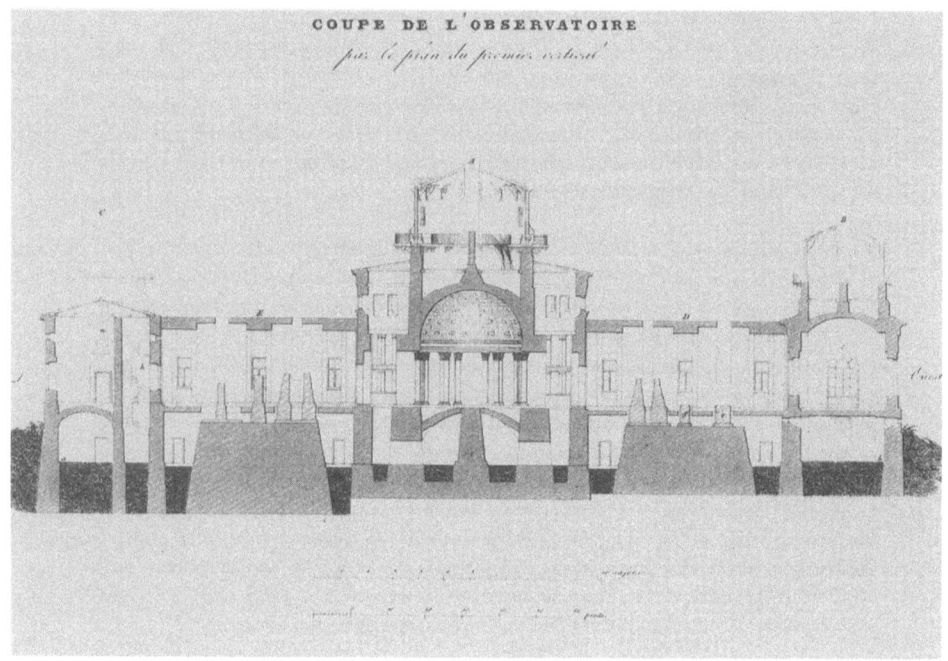

East-West cross-section of the main observatory building at Pulkovo, showing the "subterraneans" Reproduced from a copy of the "Description" at the Dominion Astrophysical Observatory.

Airy thought that there were more instruments than the staff of Pulkovo could use regularly and continuously, but recognizing that the instruments were not intended to be used thus (as were those at Greenwich) he thought the balance correct. All the instruments seemed to him "without exception, of the very first class"[10] and he repeated the story of the plans, "even the details to the smallest screws"[11] being submitted for Wilhelm's approval. Nevertheless, Airy had specific criticisms of most of the instruments. He believed the counterpoises of the transit circle ("an admirable instrument") to be too heavy

> to a point which I fear to be injurious and I know by experience that the injury which is dreaded by the German artists (namely the wearing of the pivots) does not occur even when they carry a very much greater load than is left on these [12]

While the meridian circle and the heliometer escape serious criticism, the old-fashioned mounting of the micrometers on the lesser transit circle was such as Airy "certainly should not admit in a new instrument at Greenwich" His comments on the Prime-Vertical instrument give insight into Wilhelm's observing methods:

Before I had examined this instrument at Pulkowa, I was not disposed to recognize it as one competent to the delicate determinations for which Mr. Struve intended it. After a careful examination of it, I am bound to say that my objections to it are in part removed, but not entirely. The firmness of the instrument exceeded my expectations; yet I am still unwilling to adopt for this use a form which no one for a moment would tolerate in a transit instrument. But in stating these objections to the instrument, I must state that they apply to it as an instrument to be used by astronomers in general, and not as one to be used by Mr. Struve. I had the pleasure twice of witnessing complete observations made by him; and I trust that he will not be offended by the testimony of one who, though a younger man, is not without experience as an observer, to the caution, the delicacy, the steadily waiting till the proper time, the promptitude at the proper time, which distinguish Mr. Struve's observation. In his hands I have no doubt that the accuracy of the instrument is limited only by the circumstance which he himself pointed out to me, namely the difference of radiation from the different piers; in other hands I should have no such confidence in its accuracy.[13]

Although Airy observed some close and difficult double stars with the 15-inch refractor and had "no doubt in saying that its optical part is admirable", he severely criticized the mounting:

It will hardly be credited in future ages that there was no method of nicely adjusting this celebrated telescope, either in hour angle or in polar distance, except by the hand of the observer pressing the tube or the counterpoises and thus straining the centerwork. I have no doubt that Mr. O. Struve is able to master this pressure with the utmost delicacy, and it is infinitely creditable to the education of his hand, but it is (I think) infinitely discreditable to the construction which impels him to use such an expedient. With the clockwork also I was not satisfied: it is not, I think, sufficiently powerful: its rate appears to be affected by wind etc.; and its adjustment is not sufficiently delicate.[14]

Airy was perhaps a bit harsh on Merz and Mahler, for the mounting and drive on the 15-inch incorporated many improvements suggested by Wilhelm after his experience with the Fraunhofer refractor. Others criticized the drive too, but thought the stone pier a great improvement over the wooden legs of the Dorpat telescope, that had impeded observations near the zenith. Ironically, only a few years earlier, Wilhelm had written of the remarkable ease of movement of this colossal instrument.[15] Otto felt obliged to write in "his" telescope's defence.[16]

One serious general criticism made by Airy reflects the Greenwich tradition of reducing and publishing all observations as they were made:

The most striking present deficiency is in the arrears of reductions. These, after a time, form so heavy a load, that, if I were Superintendant of the Central Observatory, I should absolutely suspend all observations till those already made had been entirely reduced and nearly printed.[17]

Wilhelm, on the other hand, deliberately adopted a different approach. He believed that observations should be published only as part of a complete investigation, as *moles bene digestae*.[18]

A corner of the Pulkovo library. (Source as for the photograph on p. 80.)

Despite his criticisms of detail, Airy was generous in his praise and careful to end on a note of commendation of "the Noble Library, which is probably the most complete in the world in reference to its peculiar subjects".[19] After Wilhelm had read Airy's account he wrote to the latter, in April, 1848:

> As to Your account of Pulkowa, I feel myself engaged to declare, that I have been in the highest manner gratified by it, and that I am proud of YOUR opinion on the Pulkova Observatory. [Wilhelm's emphasis.]

Airy returned home, laden with presents, having obviously enjoyed his visit. His journey home had not been easy, as Wilhelm's letter to him in October 1847 makes clear:

> We were very satisfied when your letter written the 28th Sept came, which gave certainty of your safe arrival at home, though after a very bad passage from Hamburgh. For a letter from Schumacher had told us before, that your steamer had to fight against a furious storm. Your stay in Pulkowa will be an object of delightfull [sic] remembrance, for a long time, to all inhabitants of the observatory. The only object I regret is that Madame Airy has not been with You. Pay to your lady my respectfull and heartfelt compliments, and tell her, if You please, that I shall never forget all the kindeness, I met with during my long though irregular stay at Greenwich.

We all wish the Samovar will be to You a quotidian remembrancer of Your Russian friends, supposing that this philosophical apparatus has got the approbation of the highest authority.[21]

The term "philosophical apparatus" had apparently been coined for the samovar by a younger member of Airy's household, who had in the past been scolded for neglecting to keep the kettle boiling. The opinion of the "highest authority" -presumably Madame Airy -- has unfortunately not been recorded. Airy was also offered a Russian knighthood by Uvarov (acting on behalf of the Tsar) namely, the second class in the order of St Stanislas. Obviously pleased, he accepted it provisionally, but --conscientious as usual-- inquired, when he got home, whether or not a servant of the British Crown might accept foreign honours. He was told that he might not, and withdrew his acceptance.[22] In due course he received a British knighthood, but curiosity led him to inquire of Wilhelm about the honour he had been offered. Wilhelm (himself the recipient of several) replied:

You ask me for a complete statement of the degrees, application and estimation of the different Russian orders of Knighthood. That is a very long object, which I am not able to explain with exactness. I think therefore it will be sufficient to notice, that the decoration of the second class of St Stanislas, which is to be borne as a neck-lace, is given by the Emperor especially to very distinguished foreign scientific men. Bessel, Schumacher, Encke, LeVerrier got it. As far as I know, the first class has been given only twice to foreigners; to Baron Berzelius and to Sir R. Murchison. Schumacher received the second Stanislas 1834, and 1845 the second class of St Anne. Humboldt is the only scientific man abroad, who bears the first class of St Anne.[23]

In the following year, 1848, a bad epidemic of cholera raged in St Petersburg and the Struve family learned one of the advantages of living outside the city, since the Observatory was relatively free of infection. Wilhelm wrote of it to Airy in a letter written at the end of the year (both Wilhelm and Otto used to catch up arrears of correspondence at the turn of the year):

In June month I have made an absence of 4 weeks from Pulkowa to pay a visit to our old residence Dorpat with Ms. Struve and 3 children, Alexandra, Olga and Charles; afterwards I went alone to Helsingfors for astronomical business, relating to some arrangements to be made in that observatory. It was just when I arrived at Helsingfors that the cholera broke out in Petersburgh with so furious an intensity, as never was known before in Europe. Near to 17000 died in two months in our Metropolis. In the surrounding country the intensity was in some degree less, though strong enough. Pulkowa observatory has been a blessed place, for not a single case of illness arrived here, when I except a general attack to which nearly all were exposed at that time. You may believe that I returned home as speedily as possible, to be at such a hard time with my family and in my place. God give that the plague does not return next year.[24]

The "general attack" seems to have been the mild form of the disease, cholerine. Otto, who "enjoyed" bad health throughout his long life, succumbed and was ill for several months. "During this feeble disposition of my body" he wrote to Airy in March 1849 "our physician... defended me strongly to strain myself by working".[25] He too commented on the epidemic:

> In my little family and in that of my father all are now in a good state of health. We
> cannot but thank the Heaven that, besides the lamented death of our little Conrad, and
> the weak state of my health, we all have passed so well through the terrible epidemics of
> the last year.[25]

Otto remarked that the early years in Pulkovo were a time of change for
the family[26]: the children of Wilhelm's first marriage were growing up, leaving
home and themselves marrying. Even the youngest, Emilie, was thirteen when
Airy visited Pulkovo. This aspect of family life is also reflected in a letter
written by Wilhelm to Airy in April 1848:

> The benevolent interest You take in all which belongs to my family, may be the excuse for
> the following particulars: The wedding of M. Döllen and my eldest daughter Charlotte has
> been the 22/10 Febr. which day is likewise my wedding day, and the birthday of Döllen's
> father, living in Berlin. Seldom a marriage can be made, with greater satisfaction to both
> families than this. -Otto will write himself to you. My second son, Conrad, who is
> physician in Kremenchug in the government of Pultawa, likewise married since January
> month. Heinrich, my third, a chemist, was in Germany, during Your visit to Pulkowa. He
> is now in Sweden, but unhappily Berzelius fell ill, and will probably not recover. He then
> intended to go to Liebig, and afterwards to Dumas at Paris. But the troublesome state of
> Germany and France forces him to abandon his projects, and to return to his father's
> house. Do you remember the tall Bernhard? He left home in January month to go to
> Irkuzk in Siberia [Wilhelm's emphasis], in advance of 4,500 miles from us. A friend of
> mine, General Muraviev, recently became Governor General of the Eastern Siberia and
> took him to his own office under very convenient conditions. You can believe that this
> was an arrangement not made by me the father, but in consequence to the wishes and
> spontaneous declaration of the young man, with which I was very satisfied, for I like a
> resolute and undertaking character. Alexandra is the support of the mother in the
> household affairs. But all adult individuals of the family are working together for the
> instruction of the little ones, for we have no public School to our disposal. Charles is going
> on very quickly in his English studies. --Even Mary, my little granddaughter [Otto's
> eldest child], comes everyday upstairs to her aunts for reading and writing.[27]

Charles, Karl or Kyrill, appears to have been privileged amongst the
children of both marriages. Otto tells us that Wilhelm would take Karl to the
office and teach him in the intervals between his own work.[28] Karl did some
astronomical work in early manhood, but became a diplomat and was, for a
while, Imperial ambassador in Washington. Heinrich, a chemist, was himself in
due course elected to the St Petersburg Academy and was apparently highly
thought of by Berzelius --the greatest chemist of his time. Wilhelm and
Berzelius had met as early as 1830 (Chapter 5) and they met again in 1833
during Wilhelm's chronometer expedition on the Baltic. Berzelius, who
presented Wilhelm on this occasion to the Crown Prince (later King Oscar I)
afterwards wrote to a friend: "He is an exceedingly pleasant spirited and
capable man, one of the greatest astronomers of our time. Cronstrand was quite
transported with happiness during his presence here, and the tears poured out
of his eyes when Struve left".[29] Thus the way was paved for Heinrich to spend
a month in 1846 with the great man, who seems to have been as impressed with
the son as he had been with the father. Writing to his pupil Wöhler, another
famous chemist, Berzelius said "he found it a great *délassement* to go and look

at his [Heinrich's] experiments and occasionally to give some advice".[30]   Plans
for a longer visit, as Wilhelm explained to Airy, came to nought.   Even
alternative plans were frustrated by Europe's year of revolutions, as Heinrich's
life echoed some of his father's early experiences.

The next visitor to leave a detailed account of a visit to Pulkovo was the
Harvard astronomer G.P. Bond.   His father, astronomer W.C. Bond, shared with
Otto an interest in the Orion nebula.   Photography of such an object being not
yet practicable, each man made drawings.   Otto criticized W.C.Bond's in terms
that both he and his son found offensive.   Nevertheless, the younger Bond
travelled to Pulkovo in August 1851 and, during his two-week stay, was warmly
welcomed by Otto.   The criticism of the elder Bond's drawing was never
withdrawn, but Otto appears to have moderated its tone sufficiently to remove
the cause of offence.   Bond sailed up the Baltic to Kronstadt and transferred to
a small launch to reach St Petersburg, where he was met by an emissary of the
Observatory.   Arriving there before noon, he immediately plunged into business.
"...In the course of the afternoon", he wrote in his diary, Wilhelm "explained to
me the method of using the great prime vertical, his favourite instrument.   In
the evening I observed, in company with M. Otto Struve, with the great
refractor."[31]   But even Bond was tired by then and went to bed, having been up
since 3 a.m.   Two days later, Sunday, he slept late but "was pleased to see here
more regard paid to the day than is usual, I think, on the continent.   Most of
the family went to Church".[32]   Later there was a large family party -a regular
feature of Pulkovo Sundays.   Bond was impressed by the fruit and vegetables
supplied from the Observatory gardens-- especially the raspberries, which also
drew praise from other visitors.

Bond was as impressed as Airy had been by Wilhelm's skill as an
observer.   "Tenths of a second of arc take the position here that seconds have
hitherto done elsewhere."[33] he wrote.   He observed with the 15-inch refractor
himself, hoping to compare it with its twin at home in Harvard, but none of
the nights he had were good enough for him to decide which was the better
optically.   After his last night he wrote "The difference is certainly not great.
My impression would be rather in favor of the Poulkova.   The purple seems not
so evident in it.   When the image is out of focus, the image is oval."[34]   Bond
was as critical as Airy had been of the drive, however:

> By practice [Otto] has acquired a facility in steadying the motion of the telescopes with his
> finger so as not to be so much incommoded by the irregularities of the clock as those less
> in practice.   The going of the Munich clocks, compared with that at the Liverpool
> Observatory is insufferably bad.   I speak as well of ours at Cambridge as of that at
> Poulkova, but the latter may be scarcely as regular as the one at Cambridge.   Both have
> great imperfections.[35]

One night, Bond and Otto observed together until 3 a.m.   They compared
their measurements of several double stars, measured Neptune's satellite and
looked at a new comet.   Bond suggested that they look at Saturn:

> O. Struve inspected it first, and I perceived instantly that he was seeing the new ring for
> the first time, and with entire certainty.   I suspected so before he spoke.[36]

The new ring was the so-called "crêpe ring" which had been discovered the year before by Bond and his father, and independently by the English amateur W.R. Dawes. Later in the morning, Bond listened to a long discussion between Wilhelm and Otto about the rings of Saturn. He was much impressed by the ease with which they could consult all the relevant books in the library. "Professor S." he wrote "is decidedly of the opinion that the ring is in the process of change, the width increasing... Indeed the fact seems beyond question, and is strong confirmation of the theory that they are in a fluid state".[37] Nothing could illustrate more clearly the difficulties of comparing and interpreting certain kinds of visual observation. The "fact" is no longer believed and the theory quite discredited.

Bond also learned a little of Wilhelm's opinion of other astronomers:

> He has a profound respect, even admiration, for the works of Sir W. Herschel. It delights me also to hear him speak of Airy and Sir J[ohn] H[erschel], and of their private characters especially. Airy knows many of the English poets by heart and Schiller also. Neither Bessel nor Hansen received a university education, and Encke did not graduate, being obliged to leave by reason of war. Hansen is regarded as the first theoretical astronomer living. Knorre of Nicolaief stands very high. Dent's clocks are preferred at Poulkova to Kessels', and his chronometers are thought to be much the best.[38]

Charles Piazzi Smyth, the colourful Astronomer Royal for Scotland whose account of Pulkovo has already been quoted several times, visited in the summer and fall of 1859. His father W.H. Smyth, son of an American-born loyalist who came to Britain after the Revolution, joined the Navy rising to the rank of Admiral, and was also an astronomer at a time when there was no clear distinction between professional and amateur. Wilhelm had met W.H.Smyth during the former's first visit to England. The Admiral also knew the Sicilian astronomer Piazzi, discoverer of the first minor planet, and named his son after this friend. Before he went to Pulkovo, Piazzi Smyth observed from the mountain tops of the Canary Islands, and became convinced that such sites would prove the best for future observatories. His colleagues were slow to agree: only in recent years, over a century after he wrote, have the Canaries been developed as an astronomical site. Later in life, Piazzi Smyth wrote on pyramidology, maintaining that the dimensions of the Great Pyramid enshrined astronomical information, known to the ancient Egyptians more accurately than to his contemporary colleagues. The resulting controversy led to his resignation from the Royal Society of London.

More mundane affairs took Piazzi Smyth to Russia. He was a practical man, who invented an early form of range-finder and was a pioneer of stereoscope photography. He also invented a gyroscopically stabilized platform for use on ships. The British Admiralty was not interested, so he decided to show it to the Russian Navy and to use the journey to test the device. To do this, he needed an introduction to one of Nicholas' younger sons, the Grand Duke Konstantin Nikolayevich, who served as minister to both his father and brother (Alexander II) and in 1859 was in charge of the Imperial Navy. As a youngster, the Grand Duke had attended a series of popular lectures given by Wilhelm in St Petersburg in the winters of 1841 and 1842. Although the lectures were in German, Wilhelm's impromptu delivery had attracted large crowds. Two years later, in the fall months, the Grand Duke came to Pulkovo from Tsarskoe Selo and "took a sort of course of observing techniques with the great instruments of the Observatory".[39]   He remained, in later life, a friend and protector of Pulkovo. Even if Piazzi Smyth had not been an astronomer himself, he would have found the Struve family useful in his mission.

As early as 1858, Otto wrote to Airy "let me know in a few words your opinion on the instrument invented by P. Smyth".[40] During that year, Wilhelm was seriously ill and Otto was Acting Director of the Observatory. When Smyth and his wife arrived, Wilhelm was convalescing in southern Europe. The visitors were met at Kronstadt by a member of the Observatory staff, and others greeted them in St Petersburg itself and conducted them to Pulkovo. It was July and the nights were short and warm. Otto complained, for example, that although he did not get to bed until 2 a.m., he was awoken at 4 a.m. by the buzzing of flies. Smyth was struck by how green the countryside appeared for July, the vegetation contrasting with the red iron roofs of the buildings. (The Observatory staff were proud of the quality of the iron, which came from the Urals.) He hardly believed that only a specially planted line of fir trees prevented snow from drifting over these roofs in winter.

Piazzi Smyth and Otto observed together with the 15-inch refractor one night. Using the garden entrance, to avoid disturbing workers on the other instruments, they found a soldier on guard at the doorway and in the dome a sergeant in uniform. Much of the work around the Observatory was performed by the soldiers, who eagerly sought positions there: a man could marry on such a posting, which became a reward for long service and good conduct. The sergeant had opened the dome and lit the lamps. He waited for Otto to set the telescope, then chose the appropriate seat and set of slow-motion handles, turned the dome (it could turn through 180° in a minute) and brought Otto the table with pens, ink and observing book. Then the sergeant stood in the dome until called. All Otto had to do was to decide what to observe, and observe it. (Oh! For such an observing assistant!) Just as Airy and Bond had been impressed by Wilhelm's observing technique. so Smyth waxes lyrical about Otto's. "Every motion told with full effect."[41] Otto was observing double stars of less than 1" separation. Smyth praised Otto's rigid, searching scrutiny, his repeated examination in slightly different positions "until, the telling moment having arrived, the micrometer screw received a dextrous final touch from thumb and finger, then, with the head thrown back at ease, the divisions were read off and duly entered in the manuscript observing book". Smyth, having looked at a pair separated by 0".3, which he saw only intermittently as double, wrote "Our friend's skill was truly something extraordinary."[41] Otto said that it was a poor night.

The quasi-military character given to Pulkovo by its staff of semi-retired soldiers was strengthened by the close association of many of the officer corps with the Observatory's geodetic work. Some officers came there for periods of geodetic training, and others were permanently attached to its staff. Otto wanted to introduce his guest to one of the latter who, he said, was Captain Smythlove. The man's name is usually transliterated "Smyslov", but Piazzi Smyth openly admitted his complete ignorance of the Cyrillic alphabet. Otto apparently enjoyed pulling his guest's leg, and would have enjoyed it even more if he ever read *Three Cities in Russia*, where Smyth repeats the pun, apparently not understanding the joke.

Otto took the Smyths on a sightseeing tour of Tsarskoe Selo Palace. He was annoyed that guide books made no mention of the Observatory -one even said that there was nothing of interest on the road after the third mile from St Petersburg, until the Palace itself! He also quarrelled with the statement that the palace was no longer used, saying "Russians are far too practical a people for such an objectless waste as that would be".[42] Otto drove the horses himself, and deliberately showed them off on the homeward journey, timing the distance of 3.5 miles in 17 minutes with his own watch.

Smyth had to stay in Russia until the Grand Duke was ready to see him. He and his wife took lodgings in St Petersburg, where Otto still visited him fairly frequently. Otto was planning a southern observing station in the Caucasus, with its headquarters in Tiflis, and was interested in Smyth's experiences in Teneriffe. Smyth was introduced to General Chodzko, who had spent the summer on Mount Ararat; a few days later Otto told his guest that the Emperor had approved the new observatory.

Wilhelm was expected back from his rest-cure on August 30 (New Style) and Otto invited the Smyths to meet him at the English Quay. It was not a family occasion, most of the astronomers from the Observatory were there. After nearly all the other passengers had left, "William von Struve" was pointed out to the Smyths, who

> "...saw a noble old man, of Scandinavian lineaments, with traces of the manly constitution of his youth eminent still, and a kindly blue eye that looked firm and true. Affectionate was the greeting that followed between the father and his long-separated children, in the number of whom, after a manner, all the other Russian astronomers seemed to desire to consider themselves and be considered. His reception of them all was dignified as well as genial, and before he drove off with them, to return to the loved abode of his labours, we had the honour also of receiving from him a friendly welcome, and in excellent English."[43]

Other visitors remarked on Wilhelm's patriarchal character: his direction of the Observatory must have been paternalistic or even autocratic. The other astronomers were discouraged from interrupting him when he was working in his office --even Otto said that he might sometimes be thrown out.[44] On the other hand, Wilhelm would take long walks with his staff in the Observatory grounds, discussing their problems, and all were welcome at social gatherings in the Director's apartment. Otto also says:

> The whole colony formed, so to speak, a single large family of which Father was the head and fulcrum. He was the honoured patriarch of the little community and Mother the refuge of all who needed help and council.[45]

Alexander von Humboldt, in his letters to Schumacher and Gauss, referred to Wilhelm as the "tyrant of Pulkovo" or even the "loud-speaking tyrant of Pulkovo".[46] (In the same place he called Otto the "Crown Prince".) These remarks were jocular --Arago said that no man had a more malicious tongue than von Humboldt, or a kinder heart.[47] There is ample evidence that von Humboldt had a real respect for Wilhelm (Chapter 5), yet the joke must have had some basis in truth to be funny at all. Wilhelm's own character was "resolute and undertaking". He drove himself relentlessly, and those around him must have been driven, if only by example. Sigurd Schulz, biographer of the artist C.A. Jensen, wrote of the latter's first portrait of Wilhelm as showing a force "close to brutality in the expression".[48] There are so many independent accounts of Wilhelm's kindness that the word "brutality" does not fit, but it does draw attention to the side of his character that might otherwise be overlooked.

After a visit to Moscow, the Smyths returned to Pulkovo and had at least one meal with Wilhelm, when he told them of the building of the Observatory. But first,

> Formally and with fervour, at his hospitable board that evening, did he propose the health of his old friend my respected father, and show all honour to the British lady my excellent wife and present companion.[49]

On September 15/7, Smyth assisted in the group observation of occultations of the Pleiades (Chapter 6). The occultations should have been observable both in Europe and America and provided an opportunity to improve the determination of longitude difference between the two continents. They occurred about a year after the first short-lived transatlantic cable had failed. Smyth lamented bitterly that the cable had not been used to transmit time signals, and thus to determine the longitude difference. Döllen had calculated the expected time and position angle of the disappearance and reappearance of each star. Each observer (whether in one of the small domes or in the grounds) was provided with a copy of this list, a small portable telescope, a chronometer and a lamp. Despite this careful preparation, the sequence of events that Smyth describes is all too familiar to any astronomer who has tried to make special observations at a definite time. Although the Sun set in a clear sky, the occultations did not begin until nearly 10 p.m. --only shortly after moonrise. The seeing began to deteriorate as the Moon approached the first star which was to disappear behind the bright limb. The star seemed to disappear and reappear several times, and accurate timing was impossible. Conditions were even worse for the second star, and then cloud covered everything. Resident observers said it was the worst seeing they had ever experienced. (It always is when there is a visiting astronomer!)

Next day, the temperature was almost down to freezing --exceptionally cold for the time of the year, said all the Struves. Smyth lost his voice --a common complaint, he was told, in the first cold weather: once winter really set in, it was less usual. He began to worry whether or not he would get home before the Baltic froze. Pulkovo prepared for winter: the soldiers exchanged uniforms for sheepskins, the astronomers and their families looked out their furs (worn with the fur inside) and all hands got the storm windows out of the subterraneans.

Soon, however, Smyth had his meeting with the Grand Duke --who was much impressed by the gyroscopic platform-- and then was able to go home on one of the last ships out of the Baltic that year. Two admirals invited by Otto had recommended the platform to the Grand Duke, who ordered the Pulkovo instrument makers to make a copy. Smyth was impressed by the Grand Duke's command of the subject and by his mastery of German --which he spoke to Otto-- and of English --which he spoke to Smyth.

Piazzi Smyth's book on Russia was published three years after his visit. The present writer read a copy 120 years after that, and had to cut the pages. At least one of Smyth's contemporaries read the book, however, and was inspired to visit Pulkovo himself. This was Cleveland Abbe (see also Chapters 4 and 6), a young American astronomer and surveyor who had worked with B.A. Gould on the U.S. Coast Survey. He volunteered for service on the Yankee side in the American Civil War, but was rejected and wrote to Otto Struve to inquire about the possibility of visiting Pulkovo. Otto replied in June 1864, suggesting that Abbe came as a "supernumerary astronomer" --a sort of unpaid post-doctoral fellow. Otto wrote:

> You will get here free lodgings with furniture, fire and light; with that the average expense of a year's residence at Pulkowa might be estimated to 300 dollars.[50]

Abbe took leave of absence and arrived in Pulkovo at the turn of 1864-5. His father provided the $300, which soon proved insufficient, for in August 1865 Abbe wrote to his parents:

> What His Excellency Otto Struve meant by writing me that one could live in Pulkovo for $300 I do not know. I guess he never tried it on.[51]

Nevertheless, Abbe stayed for two years, and is believed to have been the only American scientist who spent so long working in Russia in the nineteenth century. After his return he was, for a while, Director of the Cincinnati Observatory, but he achieved greater fame as a meteorologist in Washington. He set up the first nation-wide weather forecasting service in the United States, and was much influenced by the Struves' insistence on the practical utility of scientific research. His correspondence, including almost all the letters to and from his parents while he was in Pulkovo, is in the Library of Congress.

Abbe, brought up in a staunchly republican family, was somewhat taken aback in a land of nobles who were each to be addressed as "Your Excellency". "They had better understand" his father wrote "that we are all nobles here".[52] Abbe senior was even reluctant to address letters to his son at the "Imperial Observatory" until Cleveland complained that letters addressed to him at the "National Observatory" were being delayed. The Abbes were also rather puritanical Baptists. Unlike his compatriot Bond, Abbe did not think Sundays were well observed, and constantly complained (in his letters) of the irreligious attitudes of his colleagues. He arrived in Pulkovo at a bad time: Wilhelm had died just before, Otto himself was away convalescing in Europe. The Acting Director, Winnecke --a student of Argelander's and connected by marriage to the Struve family-- had a nervous breakdown shortly after the arrival of Abbe, who was thus left without any clear direction until Otto returned, still unwell. At first, Abbe took to Otto, but later found him "cold" --perhaps partly because of his illness and partly because of Otto's formality of manner. Abbe fell in

Cleveland Abbe. (Reproduced from Library of Congress, Cleveland Abbe Papers, Container 19.)

love with Otto's sister Emilie and even proposed marriage to her; after much hesitation, she refused, and Abbe returned home.

Abbe's letters do give a very detailed picture of everyday life in Pulkovo. He particularly liked Otto's brother-in-law, Wilhelm Döllen, whom he found "a most enthusiastic and warm-hearted man". Döllen cautioned Abbe against overwork and told him

> ...it is quite impossible for any one to live here many years alone; if he does not marry, he will overwork and sicken.[53]

There was, indeed, much sickness at Pulkovo: even the married men seemed to have overworked. Döllen himself needed a rest in Germany while Abbe was there. Nevertheless, Abbe rose at seven and went to bed at eleven and told his parents he was "buried in astronomy". Döllen again expressed horror at Abbe's timetable, but Abbe wrote:

> I never before realized how fully and wholly these Germans give themselves to study. The Americans nor any other nation can compare with them, and I feel I am in a great way behind but I have not yet acknowledged it and I hope they will not find it out.[54]

Abbe's problems were compounded by his need to learn German --the working language of the Observatory-- from scratch. If he did work, as he claimed, fourteen hours a day, he followed Wilhelm's example. Otto gives us this account of his father's typical working day in the later Pulkovo period:

> As a rule, he got up around 8 a.m., quickly drank his morning coffee and about 9 a.m. he was already at his work-desk in the Director's room of the Observatory, which he did not leave until about 2 p.m. when he was called to the mid-day meal. After that, he allowed

himself a mid-afternoon nap from 3 p.m. to 4.30 p.m., then quickly drank coffee with the family and around 5 p.m. he went back to work that was first interrupted for a little while towards 9 p.m., in order to take supper, and then continued from 10 p.m. until 3 a.m. So he spent daily 13 to 14 hours at work and scarcely stirred from his work-chair, except perhaps on a beautiful evening in summer, to go for a while in the fresh air.[55]

Abbe contrasted Otto and Wilhelm, perhaps not altogether fairly, since he never met the latter. He wrote (with his rather idiosyncratic punctuation):

I presume that those who have lived here many years find a great difference between Otto and his father for the latter was I imagine an unusually affable open-hearted gentleman. I find some of his children at least Mrs Döllen etc to be people with whom I can very well sympathize and his widow Madam Struve... I like very much indeed.[56]

At first, to judge from his letters, Abbe took to Otto --complaints of the latter's "coolness" came only later:

Otto Struve is about 47 and not a handsome man but tall and I think pleasant, at least he is to me though I find others are not much drawn out towards him, but perhaps he will change or I... I find that everyone expects that I am to stay here for many years, whether the Director has the intention to offer me any permanent position I cannot say... but I fear to stay because the society is so monotonous... also because I find a religious state of mind so foreign to the disposition of my associates.[57]

Abbe was not quite fair to the "society". He was always welcome at the Director's parties, but since they were held on Sundays, he could not attend without violating his own principles --which he sometimes did. In summer he preferred to walk to the Lutheran church in Tsarskoe Selo, sometimes in company with Victor Fuss or Wilhelm's youngest son Nikolai (Kolla) when he was home from his theological studies in Dorpat. At other times he would spend the whole week-end in St Petersburg, and take tea with Madam Struve. Hugo Gyldén, later a distinguished Swedish astronomer but then a junior researcher at Pulkovo, made Abbe welcome, as did Döllen. Abbe was much impressed by their knowledge of American affairs, their opposition to slavery and their support for the Yankee side. (This was the time of Alexander II and the liberation of the Russian serfs.) Otto did, indeed, offer Abbe a permanent appointment (Chapter 12) but the latter decided to go home, after only two years in Pulkovo, when Emilie refused his offer of marriage. He continued to correspond with Otto, and the two men met again in Washington (Chapter 14). Abbe seems to have helped to bring Simon Newcomb into correspondence with Otto, and those two men were writing to each other regularly by 1867. Otto's letters to Newcomb are also in the Library of Congress. The first few are in English, but Otto switched to German as soon as he realized that Newcomb could understand that language, for Otto --never completely at home in English-- preferred to write in German whenever he could, simply in order to write his many letters more quickly.

Newcomb visited Europe in 1870-1. He was in Paris during the time of the Commune (late spring 1871) and was in Pulkovo in March of that year.[58] He described the visit in his *Reminiscences of an Astronomer*. Newcomb travelled to Pulkovo by train: Otto had told him to telegraph from the Russian

Simon Newcomb.   (Photograph reproduced by courtesy of the U.S. Naval Observatory.)

border so that a man could meet the train at Tsarskoe Selo and bring him to Pulkovo.   Through some confusion (the railway from the frontier to St Petersburg was still relatively new) the telegram was never sent and Newcomb had to hire his own carriage --by sign language-- late on a cold winter's evening.   Nevertheless, he still wrote:

> There is no warmer welcome than a Russian one, and none in any country warmer than
> that which the visiting astronomer receives at an Observatory.   Great is the contrast
> between the winter sky of a clear moonless night and the interior of the dining room,, forty
> feet square, with a big blazing fire at one end and a table loaded with eatables in the

middle. The fact that the visitor had never before met one of his hosts detracted nothing from the warmth of his reception...

One drawback from which the astronomers suffer is the isolation of the place. The village at the foot of the hill is inhabited only by peasants, and the astronomers and employees have nearly all to be housed in the observatory buildings. There is no society but their own nearer than the capital . At the time of my visit the scientific staff was almost entirely German or Swedish by birth or language...

Once a year the lonely life of the astronomers is enlivened by a grand feast --that of the Russian New Year. One object of the great dining-room I have mentioned, the largest room, I believe, in the whole establishment was to make this feast possible. My visit took place early in March, so that I did not see the celebration; but from what I have heard, the little colony does what it can to make up for a year of ennui.[59]

Unlike Airy and Schumacher, who wrote for their colleagues, Newcomb wrote a popular book. Instead of describing the instruments in detail, he gave his impressions of the peasant village of Pulkovo. He was concerned about how long the peasants had to work and wondered when they were able to sleep. Their houses, however, he thought were probably comfortable, although meagrely furnished.

Newcomb paid a second visit to Pulkovo in 1885, taking Anita, one of his daughters with him, partly in order to inspect the new 30-inch refractor. Newcomb had persuaded Otto to have the large objective for this made in the United States (Chapter 14). Anita and her father hoped to sail up the Baltic to St Petersburg, as so many earlier visitors had done. Now there was a railway, however, Otto strongly advised against the sea voyage, either from Lübeck or even from Stockholm. It would be slow, time-consuming and somewhat unpredictable, and Otto doubted that good accommodation would be available on the steamers.[60] As a final inducement, Otto offered (again!) to meet the train with his horses, if Newcomb would telegraph from the Russian frontier. What happened this time is not recorded. Perhaps Otto tried to break his speed record, as he had done with Piazzi Smyth!

The last visitor to leave us a sketch of life at Pulkovo --right at the end of the Struve régime-- was not an astronomer but a literary man, the Vicomte Eugène Melchior de Vogüé, who, in 1889, published a vignette of the Observatory, obviously based on a recent visit he had made:

...A troika ride is the favorite amusement of St Petersburg society of winter nights, and the one that leaves in the mind of the foreigner the most vivid and novel souvenir. The inns where the Tsiganes sing are the usual object of these nocturnal excursions; sometimes, however, others are suggested; for instance to the Observatory of Pulkowa, which rises midway between Petersburg and Tsarskoe Selo, on a hill crowned with pine trees. It is the only elevation on the marshy plains which surround the capital. There lives a little German colony; for they are Germans who keep watch over the Russian heavens. With a few exceptions, this family of astronomers is recruited in the university at Dorpat, and holds its celestial fief with jealous care. When you enter Pulkowa, you find yourself transported to another world. You might imagine yourself in some calm institute in Göttingen or Jena... They live in common a patriarchal life --an honest German life, staid and serious like that of the stars.[61]

The Vicomte continues in this vein, conveying a somewhat romantic view of an astronomer's life, He is not always factually correct, but he probably conveys fairly accurately the impressions he picked up from his society friends. His emphasis, in 1889, on the Observatory's "German" character is significant (Chapter 16). In his final paragraph he tells of an aspect of Pulkovo life rarely found in the scientific histories:

> But this solitude and this peace are exposed to frequent invasions. If an eclipse is announced, the ladies of Petersburg form a party to go to Pulkowa; they either belong to the Court, or have taken the precaution to have some dignitary in their party; and consequently the Imperial Observatory could not refuse to satisfy their caprices. The troikas deposit in the temple of science the noisy visitors , who take possession of the telescopes, and demand for their particular use that corner of the heavens where something important is about to happen. The[y] have all these mysteries explained to them; they ferret about in albums of lunar photographs, and their curiosity is excited by the marvels that the old sorcerers tell them. The evening ends with supper of ham and sauerkraut prepared by Madame la Doyenne, and in listening to one of the younger German women play on the piano a sonata of Schumann or Weber. The joyous band then starts back, enchanted with the contrast between its habits of luxury and the austere simplicity of which it has just taken a glimpse.[62]

Perhaps the life at Pulkovo did seem one of "austere simplicity" to those used to the luxuries of the court, yet visiting astronomers found it ample and pleasant. Wilhelm's days of poverty were over when he arrived in Pulkovo, and Otto led a good life, with frequent visits to German spas, to take the waters. The rest of the Vicomte's tale is familiar enough. The only difference, now, is that to visit a modern observatory you need not take a court dignitary in your party; any taxpayer will do.

NOTES

1. Simon Newcomb, 1903, Reminiscences of an Astronomer, (see Chapter 6, ref. 30), p. 309.
2. H.C. Schumacher, 1840, Astronomische Nachrichten, Vol. 18, pp. 33-44 (p. 33), quoted with permission.
3. Simon Newcomb 1903, Reminiscences of an Astronomer, (ref. 1), p. 291 reports a similar remark that he ascribes to Airy's contemporary, the mathematician Sir William Rowan Hamilton.
4. F.G.W. Struve, 1841, letter of September 14 to G.B Airy (RGO 939 f.529). This and all correspondence between Airy and the Struves is reproduced by permission of the Royal Greenwich Observatory.
5. F.G.W. Struve, 1847, letter of January 23/February 4 to G.B. Airy. (RGO 939 f. 555).
6. O.W. Struve, 1847, letter of February 12 to G.B. Airy (RGO 939 f. 571).
7. G.B. Airy, 1848, Astronomische Nachrichten, Vol. 26, pp. 353-360 (p.354). This and following extracts quoted with permission.
8. G.B. Airy, 1848, ibid., p. 355.
9. G.B. Airy, 1848, ibid., pp. 356-7
10. G.B. Airy, 1848, ibid., p. 357
11. G.B. Airy, 1848, ibid., p. 357
12. G.B. Airy, 1848, ibid., p. 357
13. G.B. Airy, 1848, ibid., pp. 358-9
14. G.B. Airy, 1848, ibid., p. 359
15. F.G.W. Struve, 1844, Astronomische Nachrichten, Vol. 19, pp. 281-3 (p. 282).

16   O W Struve, 1848, <u>Astronomische Nachrichten</u>, Vol 27, pp 67-70   A further brief reply from Airy
      follows this paper
17   G B Airy, 1848, <u>Astronomische Nachrichten</u>, Vol 26, pp 353-360, (p 360)
18   O W Struve, 1895, <u>Erinnerung</u>, p 39
19   G B Airy, 1848, see ref 17
20   F G W Struve, 1848, letter of April 20/8 to G B Airy (RGO 939 ff 567-8)
21   F G W Struve, 1847, letter of October 22/10 to G B Airy (RGO 939 f 561)
22   Wilfred Airy, (ed), <u>Autobiography of Sir George Biddell Airy, K C B</u>   Cambridge University Press, p
      188-193   (Little more than a collection of notes for an autobiography and published posthumously)
23   F G W Struve, 1848, see ref 21
24   F G W Struve, 1848, letter of December 31/19 to G B Airy (RGO 940 ff 168-9)
25   O W Struve, 1849, letter of 27 March to G B Airy (RGO 941 ff 209-210)
26   O W Struve, 1895, <u>Erinnerung</u>, p 55
27   F G W Struve, 1848, see ref 20
28   O W Struve, 1895, <u>Erinnerung</u>, p 57
29   J J Berzelius, 1833, letter known to me only through a translated excerpt provided by Nils Lindhagen
30   J J Berzelius, 1846, letter to F Wohler, see ref 29
31   E S Holden (ed), 1897, Memorials of W C and G P Bond, C A Murdock and Co , San Francisco and
      Lemke and Buechner, New York City   Reprinted 1980, by Arno Press, New York, p 97
32   E S Holden, 1897, ibid , p 98
33   E S Holden, 1897, ibid , p 100
34   E S Holden, 1897, ibid , p 104
35   E S Holden, 1897, ibid , p 99
36   E S Holden, 1897, ibid , p 102
37   E S Holden, 1897, ibid , pp 102-3
38   E S Holden, 1897, ibid , p 101
39   O W Struve, 1895, <u>Erinnerung</u>, p 59
40   O W Struve, 1849, letter of March 1 to G B Airy (RGO 948 f 503)
41   C P Smyth, 1862, <u>Three Cities in Russia</u> (see Chap 4, ref 8) Vol 1, p 142
42   C P Smyth, 1862, <u>ibid</u>, Vol 1, Chap 11
43   C P Smyth, 1862, <u>ibid</u>, Vol 1, p 335
44   O W Struve, 1895, <u>Erinnerung</u>, p 61
45   O W Struve, 1895, <u>ibid</u>, p 55
46   K -R Biermann (ed), 1979, <u>Briefwechsel zwischen Alexander von Humboldt und Heinrich Christian
      Schumacher</u> (see Chap 3, ref 9), pp 116, 122, 136
47   L Kellner, 1963, <u>Alexander von Humboldt</u>, (see Chapter 5, ref 17), p 1
48   Sigurd Schultz, 1932, <u>C A Jensen</u>, Copenhagen (two vols)   The work is known to me only from brief
      translated excerpts provided by Nils Lindhagen
49   C P Smyth, 1862, <u>Three Cities in Russia</u> (see Chap 4, ref 8) Vol 2, p 161
50   O W Struve, 1864, letter of June 25 to Cleveland Abbe (Library of Congress, Cleveland Abbe Papers,
      Container 2)
51   Cleveland Abbe, 1865, letter of August 1 to his mother (<u>loc cit</u>)
52   Abbe senior, 1864, letter to his son Cleveland (<u>loc cit</u>)   See also N Reingold "A Good Place to Study
      Astronomy" <u>Library of Congress Quarterly Journal of Current Acquisitions</u>, Vol 20, pp 211-217, 1963
53   Cleveland Abbe, 1865, letter of January 2 and 4 (N S) to his parents (Library of Congress, Cleveland
      Abbe Papers, Container 2)
54   Cleveland Abbe, 1865, letter of January 17 to his parents (<u>loc cit</u>)
55   O W Struve, 1895, <u>Erinnerung</u>, p 63
56   Cleveland Abbe, 1865, letter of October 1 to his parents (Library of Congress, Cleveland Abbe Papers,
      Container 2)
57   Cleveland Abbe, 1865, <u>ibid</u>

58.   Newcomb does not give the exact date in his <u>Reminiscences</u> (see Chapter 6, ref. 30); it is mentioned by O.W. Struve in a letter of March 16, 1872 to Newcomb. (Library of Congress, Simon Newcomb papers, Container 40).

59.   Simon Newcomb, 1903, <u>Reminiscences of an Astronomer</u>, (see Chap. 6, ref. 30), pp. 312-3).

60.   O.W. Struve, 1885, letters of February 26 and July 8 to Simon Newcomb (Library of Congress, Simon Newcomb papers, Container 40).

61.   E. Melchior de Vogüé, 1889, <u>Harper's Magazine</u>, vol. 78, p. 851; reprinted in <u>Publications of the Astronomical Society of the Pacific</u>, Vol. 7, pp. 285-7, 1895, with whose permission it is reprinted.

62.   E. Melchior de Vogüé, 1889, <u>ibid</u>., p. 287.

CHAPTER 8

MEASURING THE SKY

Wilhelm took at least one unfinished piece of business from Dorpat to Pulkovo --the measurement of parallax. It was natural that, as one of the best observers of his time, with access to some of the best instruments, he should be in the forefront of those attempting to meet this supreme challenge to early nineteenth-century astronomy (Chapter 3). Much of his effort in Dorpat had been devoted to this, but complete success eluded him until the very end of his time there. Then, in little more than a year --between the end of 1838 and the beginning of 1840-- three men published values for the parallaxes of three different stars, that convinced astronomers throughout the world in a way that so many earlier investigators had failed to do. Wilhelm was one of those three; Bessel and Thomas Henderson were the other two. Their independent and nearly simultaneous success was not just coincidence. Many astronomers were striving for this goal, the achievement of which depended on the availability of sufficiently accurate instruments. As Europe recovered from the Napoleonic wars, such instruments were made --especially by Fraunhofer, who made those used by both Wilhelm and Bessel in their successful measurements.

The triumph represented by these measurements can perhaps best be conveyed by a description of the many corrections that must be made to the observed position of a star before its true position can be recorded. We live at the bottom of an ocean of air and, until very recently, could observe the stars only through it. One of its effects on starlight is to bend the rays so that any star appears to us to be closer to the zenith than it is. The amount of its atmospheric refraction depends on the temperature and pressure of the air, but even more on the distance of the star from the zenith. At the zenith itself, there is no displacement at all --which is one reason why Bradley tried to measure the parallax of a star passing nearly overhead (Chapter 3). At the horizon, on the other hand, a star is displaced upwards by more than the apparent diameter of the Sun or Moon. The aberration of light, discovered by Bradley, also affects the direction in which the star is seen in the sky, and by an amount much larger than the expected displacement caused by parallax. In addition, all measurements of the positions of stars are referred to moving zero points. Bound as we are --or were until a few decades ago-- to a moving earth, our imaginary coordinate systems used to map the sky are related to the Earth, particularly to the axis about which it rotates. This axis moves in space in a motion called *precession*, known to Hipparchos (2nd century B.C.) but not explained until the time of Newton. It is further complicated by a nodding motion, *nutation*, also discovered by Bradley, related to the motion of the Moon. Parallax can be determined only from observations that extend over at least a year (usually several years), and precession and nutation together can introduce much larger changes in a star's apparent position, over this time, than does

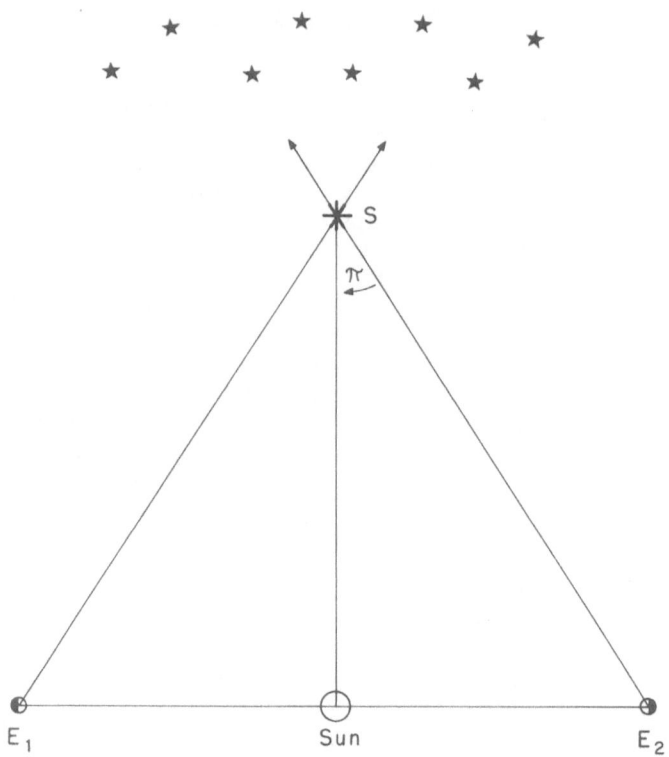

The positions of the Earth at two times, six months apart, are shown at E_1 and E_2. A relatively nearby star, S, appears to change its position, with respect to the more distant background stars, as the Earth moves around the Sun. In modern terms, the angle π is called the annual parallax of the star. For all known stars, π is less than one second of arc. (Not to scale.)

parallax itself. Finally, each star moves through space relative to the Sun, and this may change the direction in which we see a star. This change of direction, the star's *proper motion*, first recognized by Halley in 1718, may again be much larger than parallax --indeed, one criterion for selecting stars for parallax determination is that they should have large proper motions --then they are likely to be nearby.

The astronomer who would determine parallax must not only measure the very small parallactic displacement (always less than 1"); he must show convincingly that that displacement is there after allowing for all those other effects and removing them from his raw measurements. The wonder would be that *any* parallaxes at all are measured, if it were not that all causes of changes to the apparent positions of stars --except proper motion and parallax itself -- depend only on those positions. The *difference* between refraction or precession for two stars in the same field of view of a telescope is very much *smaller* than the parallaxes of nearby stars. Thus, all parallaxes are measured differentially, whether with respect to one companion, close in the sky but believed to be much further from us, by the method that Galileo proposed, Herschel attempted and Wilhelm successfully executed, or with respect to several background stars --also assumed to be much farther away-- in the manner pioneered by Bessel and which has become routine. Of course, even the fainter background stars have some parallax, and modern astronomers have devised statistical methods of allowing for this. At the beginning, however, Bessel and Wilhelm could only assume that the fainter stars had no parallax at all.

Any scientist who wishes to convince his colleagues that he has successfully measured a small quantity must not only tell them the value he obtained, but also give them some estimate of the range of values that could still satisfy his observations. There are statistical rules for deriving this range, which are relatively simple, provided certain assumptions are made about the nature of errors of observation. Wilhelm usually quoted the so-called *probable error* --a rather misleading term since, as has often been observed, the quantity itself is neither probable nor an error. The term does have a quite precise meaning in certain well-defined circumstances. Here, it is sufficient to remark that a parallax less than three times its probable error will usually not be considered reliably determined. In Wilhelm's day, the criterion was perhaps not applied quite so strictly, but he and his contemporaries were well aware that the ratio of a parallax to its probable error should be as large as possible-- a fact important for estimating the worth of some of the early attempts to determine parallax.

Wilhelm began to think about the problems of measuring parallax as soon as he became a Dorpat professor. Bessel, nine years older, made his first attempts to measure the parallax of 61 Cygni about the time that Wilhelm graduated. Although Bradley's careful work in the eighteenth century had set an upper limit of 1", or at most 2", to parallaxes, later observers claimed to have found larger values that are now recognized as spurious. Bessel examined Bradley's work again and showed that the brightest stars (assumed to be amongst the nearest) must have parallaxes less than 1". Several people recognized that the large proper motion of 61 Cygni probably meant that the star was nearby, and Bessel published two values for its parallax in 1815 and 1816. Both values were negative --a geometric impossibility, at least in a Euclidean universe-- but they did serve to indicate how small the parallax must be (other observers had found very uncertain values of about 0".5). Wilhelm's first attempts to determine parallax were made in 1818-21, before he had Fraunhofer's refractor. He observed a number of circumpolar stars with the meridian circle as they crossed the meridian, both above the pole and below it (in the northern sky). The times of their meridian passage would be affected by parallax, but not by refraction, and the nearer a star is to the pole the greater the effect. Wilhelm combined stars in pairs whose members came to the meridian about 12 hours apart and, in this way, measured the sum of two parallaxes. He showed that the average parallax of bright stars circumpolar at Dorpat was less than 0".1. He even published an individual parallax of 0".163 (with a probable error of 0"026) for δ Ursae Minoris.[1] Despite its small probable error, this value apparently did not carry conviction and is, indeed, some forty times larger than the modern (still unreliable) value. Although this

STELLARUM DUPLICIUM ET MULTIPLICIUM

MENSURAE MICROMETRICAE

PER MAGNUM FRAUNHOFERI TUBUM

ANNIS A 1824 AD 1837

IN SPECULA DORPATENSI

INSTITUTAE

ADJECTA EST SYNOPSIS OBSERVATIONUM DE STELLIS COMPOSITIS DORPATI

ANNIS 1814 AD 1824 PER MINORA INSTRUMENTA PERFECTARUM,

AUCTORE

F. G. W. STRUVE,

A CONSILIIS STATUS ACTUALIBUS, ORDINIS ST. ANNAE SECUNDAE CLASSIS CORONA DECORATI ET ORDINIS DANEBROGICI EQUITE, ACADEMIAE
SCIENTIARUM CAESAREAE PETROPOLITANAE MEMBRO ORDINARIO, IN UNIVERSITATE DORPATENSI ASTRONOMIAE PROFESSORE ET SPECULAE
DIRECTORE, SOCIETATUM REGIARUM LONDINENSIS, ASTRONOMICAE LONDINENSIS, HAFNIENSIS, GOTTINGENSIS HABLEMIENSIS, EDINBURGENSIS,
ACADEMIARUM SUECICAE HOLMIENSIS, AMERICANAE BOSTONIENSIS, SOCIETATUM NATURAE SCRUTATORUM MOSQUENSIS, LITERARIAE MITAVIENSIS,
MATHEMATICAE HAMBURGENSIS ET OECONOMICAE LIVONICAE, AUT MEMBRO AUT SODALI, INSTITUTI FRANCOGALLICI, ACADEMIARUM
REGIAE BEROLINENSIS ET PANORMITANAE A COMMERCIO LITERARIO.

EDITAE JUSSU ET EXPENSIS ACADEMIAE SCIENTIARUM CAESAREAE PETROPOLITANAE.

PETROPOLI,

EX TYPOGRAPHIA ACADEMICA.

1837.

Title page of Mensurae Micrometricae. *(Courtesy of Swarthmore College Library.)*

early work was considered an appreciable advance at the time, Wilhelm himself later criticized it severely,[2] believing that the meridian circle could not remain stable in the interval between upper and lower transits of the same star, and he decided that progress could be made only by adopting the differential method first proposed by Galileo.

The arrival of the Great Refractor and Fraunhofer's micrometers gave Wilhelm his first opportunity to try this new method. Even by 1827 he had selected Vega as a suitable star on account of its "companion", 43" away, that he correctly believed to be much more distant (Chapter 5). He argued that Vega, being bright, was likely to be near. It is also high in the Dorpat sky, when close to the meridian, and therefore easier to observe. In the *Introduction* to the *Mensurae Micrometricae* (published in 1837), Wilhelm wrote that in that same year, 1827, it was his "intention to enter on the subject of annual parallax, but [he] was called aside from it by [his] labours on double stars".[3] Even so, in that year he made preliminary measures of the interval between the transit of Vega and that of its companion (and of other similar pairs) and found that he could determine the interval with a probable error of 0".52. Thus, he deduced, if he made enough measures (the probable error decreases in proportion as the square-root of the number of measures increases) "a parallax of from 0".1 to 0".2 would not ultimately escape".[4]

This section of the *Introduction* of the *Mensurae Micrometricae* in which Wilhelm discusses parallax well illustrates the development of his thought since the publication of the *Catalogus Novus*. In the new book, he wrote of the criteria by which stars should be selected for parallax measurement. He still considered the brighter stars to be, on average, nearer, but he had clearly progressed beyond regarding all stars as of the same brightness, for he writes

> ...the brighter stars are in general nearer to the Earth... the greatest parallaxes are to be looked for in stars of the first order [magnitude], though in particular cases it may happen, that a greater parallax belongs to a star of less brightness.[5]

Wilhelm then cited a calculation by Steinheil who estimated that, if the Sun and Arcturus are equally luminous, the former must be moved over three million times farther away than it is now, in order for it to look the same brightness as Arcturus. From this, a parallax of 0".063 can be deduced for Arcturus. It is notoriously difficult to compare the brightness of a bright extended source like the Sun with that of an apparently fainter point source. Steinheil did not do too badly, considering the instruments he had, but he did overestimate the ratio of the brightnesses by about 200 times. On the other hand, if the Sun and Arcturus really were at the same distance from us, Arcturus would look nearly 100 times brighter than the Sun. The two errors tend to cancel, and Wilhelm's estimate of the parallax of Arcturus was not far from the modern value of 0".097. Wilhelm also deduced from this calculation that the apparent diameters of the images that telescopes form of stars must be spurious. It was left to his friend Airy to show that, in perfect seeing, the aperture of the telescope determines the apparent diameter of the image.

Wilhelm's second criterion for selecting a star for parallax measurement was precisely that already used by Bessel, and others, to pick 61 Cygni:

> Another indication of the distance of the fixed stars is to be found in their proper motions. Whatever be the origin of this change of place, its effects must be in the inverse ratio of the distance of the star from the solar system.

If we imagine 61 Cygni to be placed at half its present distance from the Sun, its absolute motion remaining unchanged, the apparent annual motion will be 10"332 instead of 5"166 Hence it follows that, in general, those stars must be considered nearest to us in which the greatest proper motion exists [6]

Wilhelm had remarked on the large proper motion of 61 Cygni in the *Catalogus Novus*, using it as an argument to support his belief that the two stars were related. He did not say there explicitly that the large proper motion was a likely indication that the pair was nearby. Perhaps this was because he was looking for pairs like Vega, with a faint optical companion that he could measure. The components of 61 Cygni had no convenient companions that Wilhelm could have measured with his micrometers.

Wilhelm's

third criterion for judging of the distances exists in double stars themselves   In p Ophiuchi* according to Sir J Herschel, the apparent semi-axis major = 4"3   At half the distance it would appear to be 8"6   Those double stars, therefore, must be considered nearest to us in which the angular motion is greatest   But the differences of their linear dimensions are indicated by the periods of their revolutions If we then call the parallax of a double star π , the sum of the masses $=M$ (the mass of the Sun being 1, and the radius of the Earth's orbit 1), and if from observations we should find (a) the seconds, subtended at the Sun by the semi-axis of the true orbit, and T the time of revolution

Then
$$\pi = \frac{a}{T^{2/3} \qquad M^{1/3}}$$

If we apply this formula to those double stars whose orbits are best known, from the elements of Sir John Herschel, for example, for three stars, and of Maedler for σ Coronae and ξ Ursae Majoris, we have for

γ Virginis	$\pi^3\sqrt{M} = 0"185$	(0"099)**
Castor	$= 0"200$	(0"067)
σ Coronae	$= 0"085$	(0"045)
ξ Ursae Majoris	$= 0"147$	(0"137)
p Ophiuchi	$= 0"236$	(0"202)

to which we may add, if we estimate for   Herculis $=1"0$ and $T=14$, and for 61 Cygni $a =16"$ and $T=500$

ζ Herculis	$\pi^3\sqrt{M} = 0"172$	(0"102)
61 Cygni	$= 0"254$	(0"294)

---

*   Known today as 70 Ophiuchi, or sometimes 70 p Ophiuchi
**   For comparison, modern values of the parallaxes of these stars have been added in parentheses (taken from the <u>Catalogue of Bright Stars</u>, Yale University Press, 1982)

We see from these examples that the parallaxes of double stars may amount to 1/4", if their masses are not greater than that of the Sun. For 61 Cygni the parallax would be 0".1, if the sum of the masses were even 16.4. It is, indeed, worthy of attention, that the stars in which there are the greatest proper motions, viz. 61 Cygni , p Ophiuchi and ξ Ursae Majoris, offer also the greatest values of $\pi \sqrt[3]{M}$. And in this estimation of parallaxes, deduced from the periods and semi-axes of the orbits of double stars, no less confidence is to be placed than in that which is deduced from photometrical considerations. In the one case we suppose the brightness of the Sun and of a fixed star to be the same at the same distance; and in the other their masses are to be equal. Both hypotheses must be taken as equally probable, though, in a particular case, both may be very far from the truth.[7]

In these last few sentences, Wilhelm displays considerable physical insight, but he could not have known that his argument from the masses of stars was much stronger than that from their luminosities. There is a very large range indeed in stellar luminosities, but almost all stellar masses lie between one-tenth and ten times that of the Sun. Stars very much more massive than the Sun are very rare and there are none nearby. Stars much less massive are intrinsically faint and, even if they were near us, would be unlikely to have attracted the attention of those trying to measure the first parallaxes. Moreover, since the value deduced for the parallax depends on the cube-root of the mass, it did not matter very much that Wilhelm did not know any stellar masses --as he pointed out for 61 Cygni. Since virtually all stars have masses within a factor of 100 of each other, the cube-roots will be within a factor of four or five. Similarly, the uncertainty in some of the orbital periods that Wilhelm assumed is not very important. This is why all the parallaxes that Wilhelm deduced were of the right size. In this section of the *Mensurae Micrometricae*, Wilhelm laid the foundation for the determination of what are now called *dynamic parallaxes*. By this means, a useful estimate can be made of the distance of a binary star, even if it is too far away for its parallax to be measured directly. We now know that Bessel anticipated Wilhelm in this approach too, in a letter written to Olbers in 1812.[8] Whether Wilhelm had the same idea independently, or derived it from conversations with Bessel, probably cannot now be determined. Either hypothesis is plausible. We know enough about stellar masses now to be able to make a good guess of the likely masses of binary components, so we can use the method even more effectively than Wilhelm did. Sometimes we can go even further, and use measurements of the orbital speeds, in the line of sight, of the two component stars to determine the real size of the orbit and therefore the parallax. This was done recently for 70 Ophiuchi itself, and the result agrees well with the directly determined trigonometrical parallax.[9] Wilhelm would have been delighted with this, if he had lived to see it --such measurements were not possible until about the time that Otto retired-- but he sounded a note of caution, as valid now (and as often overlooked) as it was then:

Very famous men --Savarius, Encke, the younger Herschel especially, and finally Mädler-- have devoted time and effort to calculating from observations the true orbits of double stars; and from their labours the approximate times of revolution and the other elements of certain systems --one true diameter alone being excepted because of a lacking parallax- - are generally known. But in all these investigations I find it assumed that the motions take place according to the laws of Kepler, a fact which is first to be determined by the observations themselves.[10]

This remark introduced a discussion of errors and uncertainties in Herschel's double-star measurements, but it draws attention to a circularity of argument of which we are still guilty. We sometimes point to the orbits of visual double stars as evidence for the universality of Newton's law of gravitation, of which Kepler's laws are consequences, yet the errors of the observations (including Wilhelm's excellent ones, but possibly excluding some of the best modern interferometric ones) would still mask deviations from Kepler's laws much greater than those that would arise from substituting Einstein's law of gravitation for Newton's.

After laying down these criteria for stars likely to show a measurable parallax, Wilhelm then listed stars that he considered to meet the criteria. He emphasized that stars that met all three, or even any two of them, were particularly worthy of attention. For example, 70 p Ophiuchi satisfies the second and third criteria (even the first, to the extent that it is easily visible to the naked eye on a clear moonless night) and is, indeed, one of our nearest neighbours. Similarly, Arcturus satisfies the first two criteria; although we now know that its brilliance is partly an intrinsic property, and the star -- though not at a very great distance-- is not one of the closest. Wilhelm listed seventeen stars that satisfied at least two of his criteria. Many years later, Jackson pointed out that seven stars of the 19 listed by Eddington as then being known to have parallaxes greater than $0.''2$ were to be found in Wilhelm's list, as were six out of 27 known to have parallaxes between $0.''1$ and $0.''2$.[11] Both α Lyrae (Vega), whose parallax Wilhelm measured, and 61 Cygni, which Bessel measured, were amongst those seventeen stars.

Having established the likely size of parallaxes, Wilhelm then discussed methods of measuring them. It was at this point that he dismissed his own work of 1818-21 and stated:

> The only method therefore, which in the author's opinion, can lead to the detection of parallax consists in comparing the star (A), in which a greater parallax is expected, with the stars (B) lying in its neighbourhood, which have probably no parallax: for example, a bright star with a remarkable proper motion, with small [i.e. faint] adjacent stars that have no proper motion... In this place the author only speaks of those observations which have been made with a wire-micrometer attached to his equatoreally-mounted Fraunhofer telescope.[12]

Not all the likely stars, however, have nearby faint companions to which measures can be referred. Sirius is an example (Wilhelm wrote before the discovery of its companion), but there are two stars at almost the same declination, one crossing the meridian 1^m 52^s before and the other 32^s after Sirius. Wilhelm suggested that these three transits be frequently timed at different seasons --without moving the telescope between observations. He proposed using three clocks, each beating different fractions of a second, so that the transit times could be estimated very accurately. Sirius, being closer than Vega, would have been a good star to measure for an observer in the southern hemisphere. It never rises very high in the Dorpat sky, and is, of course, unobservable from there during much of the year. Had Wilhelm been able to adopt his own proposal, he might have discovered the irregularities in the motion of Sirius (caused by its faint companion) before his friend Bessel.

Following the decision quoted above about the best way to proceed, Wilhelm did not repeat his 1827 work on timing the transits of Vega and its companion, but made a series of 17 micrometer measurements in 1835-6. The climax of the section devoted to parallax in the *Mensurae Micrometricae* is the summary of the results of these measures. From them, Wilhelm determined three quantities --the change in mean separation of the two stars over the interval of observation, the change in their mean orientation, and the parallax of Vega, which he found to be 0."125 with a probable error of 0."055. He concluded that these

> ...results prove the parallax of α L′yrae to be very small indeed, since probability confines it between 0."07 and 0."18. This cannot, however, be considered as finally settled. A continuation of the observations, through all the periods of the parallax will confine the uncertainty within narrower limits. In the meanwhile this, at least, seems free from doubt, that the parallaxes of stars not less than 0."1 may be detected with [this] instrument, by the use of [this] method...[13]

Wilhelm had begun his observations of Vega in 1835 and continued them, and prepared the *Mensurae Micrometricae* for publication, just at the time he was planning the building of Pulkovo Observatory. Nevertheless, he undertook this series of delicate measurements which, of necessity, had to be made over several years *with the same instrument*, if they were to have any value at all. Perhaps if he had not been involved in so many other things, Wilhelm would have been, beyond question, the first to measure a stellar parallax definitively. Bessel, however, was almost equally preoccupied and might well have succeeded earlier with the double star 61 Cygni, if *he* had been less busy. The value of 0."125 for the parallax of Vega happens to be close to the modern value, but it was not accepted as final and it is clear that Wilhelm did not claim that it was. It did offer hope that soon someone would measure a parallax. Bessel himself wrote to Olbers, in October 1837:

> I think Struve has taken the lead, for he has made an attempt which, though not yet a complete success, nevertheless seems to offer good prospects.[14]

Bessel had begun his successful series of observations before Wilhelm had started to measure Vega, but Bessel published nothing until October 1838, over a year after the appearance of the *Mensurae Micrometricae*, when he sent a complete account to Schumacher and a summarizing letter to Sir John Herschel. Schumacher published the paper in the *Astronomische Nachrichten* before the end of that year, and Herschel --through whom Bessel had chosen to communicate with British Astronomers because, as he said, "I can write in my own language, and thus secure my meaning from indistinctness"[15]-- promptly translated and published the letter. In it, Bessel wrote:

> I began the comparisons of this star [i.e. 61 Cygni] in September 1834, by measuring the distance from two small stars of the 11th magnitude, of which one precedes and the other is to the northward. But I soon perceived that the atmosphere was seldom sufficiently favourable to allow of the observation of stars so small; and, therefore, I resolved to select brighter ones, although somewhat more distant. In the year 1835, researches on the length of the pendulum at Berlin took me away for three months from the observatory; and when I returned, Halley's Comet had made its appearance and claimed all the clear nights. In

Friedrich Wilhelm Bessel: copy by the artist of the portrait painted by Jensen for Pulkovo Observatory   Argelander praised the likeness of the portrait to Bessel   The copy was painted for Bonn Observatory and is reproduced with the permission of the Universitäts-Sternwarte, Bonn.

1836 I was too much occupied with the calculations of the measurement of a degree in this country, and with editing my work on the subject, to be able to prosecute the observations of α Cygni [sic, but presumably 61 Cygni is meant] so uninterruptedly as was necessary, in my opinion, in order that they might afford an unequivocal result. But in 1837 these obstacles were removed, and I, thereupon, resumed the hope that I should be led to the same result which Struve grounded on his observations of α Lyrae, by similar observations of 61 Cygni.[16]

Taken together, Bessel's letters to Olbers and Herschel indicate that he did not fully accept Wilhelm's result but that it encouraged and stimulated his own efforts. He used a 6-inch heliometer, an instrument he designed himself and which Fraunhofer built for him. As its name implies, it was originally intended for measuring the angular diameter of the Sun, but it could be used for other measurements. After Bessel's success, heliometers were used for parallax measurements, but no rapid progress was made in that work until photography could be applied to it, making the heliometer obsolete. Bessel's instrument was an ordinary refracting telescope with its object-glass cut along a diameter. The two halves could be moved with respect to each other along their common diameter, and they could also be rotated around the optical axis of the telescope. Each half formed an image of the star and the two images coincided when the two halves were perfectly matched to look like an uncut lens. If two stars were in the field of view, the object-glass could be rotated so that the image of each lay on the division between the two halves of the lens, and then one half could be displaced along the dividing line until the image of one star formed by that half coincided with the image of the other formed by the other half. The displacement of the two halves was a measure of the separation between the stars in the plane of the sky. This instrument gave Bessel an advantage over Wilhelm since, with it, one could measure accurately stars of greater separation than with the micrometer. Bessel eventually chose as reference stars for 61 Cygni "two between 9th and 10th magnitudes; of which one a, is nearly perpendicular to the line of direction of the double star; the other, b, nearly in this direction".[17] The first was over 460" from the pair 61 Cygni and the second over 700". Wilhelm could not have measured these separations with his instrument and was thus constrained to work on the less promising star, Vega. From his 1837 observations, Bessel deduced an annual parallax of 0".3136 for the two stars in 61 Cygni, with a probable error of 0".0141 --small enough to convince his colleagues that he had indeed measured the parallax of a star.

Bessel's paper was published in the December 1838 number of the *Astronomische Nachrichten*.[18] In the following month, Thomas Henderson published a value for the parallax of α Centauri, based on observations he had made at the Cape of Good Hope in 1832-3. The star meets all three of Wilhelm's criteria and, with its faint companion, is still the Sun's nearest known neighbour. Henderson did not originally intend to determine the parallax from his observations, which were made with an instrument distinctly inferior to those constructed by Fraunhofer and notorious for its systematic errors. Only after he returned home to Scotland did Henderson examine his measurements of α Centauri to see if he could determine the parallax of that star. Cautiously, remembering the reputation of the instrument he had used, Henderson waited for confirmatory observations. He derived a parallax close to 1" and more than

ten times   its probable error[19] (the modern value is 0".75), but his caution --
admirable from a strictly scientific point of view-- cost him any claim to
priority in the measurement of stellar parallax.

In the meantime, despite his increasing involvement in the building of
Pulkovo, Wilhelm tried to improve his result.   In the very year that he
commissioned the instruments for the new observatory, arranged its opening
and moved himself and his household from Dorpat, he still found time to
reduce and to publish his new measure of Vega. In October 1839, just one year
after the completion of Bessel's paper, Wilhelm sent to Schumacher another
paper containing a revised estimate of the parallax of Vega, which was
published in the *Astronomische Nachrichten* early in 1840:

> In my Mensuris micrometricis stellarum duplicium etc. p. CLIX, already at the beginning
> of 1837 I had, from 17 measures of the distance between α Lyrae and the small star
> separated from it by 43", undertaken with the Dorpat refractor, made the attempt to
> determine the parallax of α Lyrae, with that of the small [star] supposed =0. The hope
> expressed there that continued observations of that kind would soon give to know the
> parallax with greater certainty is now fulfilled.   In an Addendum to the Mensuris
> micrometricis, in which micrometer measures of compound stars, up to my departure from
> Dorpat Observatory are continued, and which I submitted to the Imperial Academy on 9
> Oct/27 Sept of this year [1839], the measures of α Lyrae go up to August 1838, so that
> the mutual position of the two stars was determined micrometrically 96 times.   The
> circumstance, that for the success of these measures a combination of the greatest
> transparency of the atmosphere with the complete stillness of the images was required,
> explains why the number of them was not greater.
>
> From these measures now the parallax can be derived in two ways, namely as well from
> the observed distances, as from the measured directions of the line joining the two stars
> referred to the parallel of declination, the so-called position angles. Since, however, there
> are circumstances that impair the certainty of these latter, so they may not be taken into
> account for the determination of parallax, and it was necessary to derive this from the
> distances alone.[20]

From these 96 observations, Wilhelm determined the mean separation of
the two stars in the autumn of 1837, the proper motion of Vega itself, and the
parallax of 0".2613 with a probable error of 0".0254. He concluded:

> Since the value found for the parallax is more than ten times as large as its probable error,
> since no source of systematic error acting on the parallax is supposed, in that particularly
> the influence of temperature on the value of a revolution of the screw is determined with
> such accuracy, that for 43" distance even for extreme temperatures no relative uncertainty
> of 0".001 is found, so it appears to me that no residual ground remains, to bring the
> parallax found into doubt, and I state on the strength of it the distance of the star α
> Lyrae from the solar system equals 771,400 semi-diameters of the Earth's orbit, through
> which space light travels in 12.08 years.[21]

Wilhelm must have slipped in his arithmetic, since the figures he gives
are not quite consistent with each other. Like Bessel, he expressed the distance
of the star in terms of the mean distance from Earth to Sun --a unit whose size
was still not accurately known-- or in terms of the time that light takes to

travel the distance. This graphic way of describing the vast distances of stars, still used in popular texts, was apparently introduced as early as 1694 by an author who argued that the stars must be at least six light-weeks away.[22]

Bessel, too, revised his original estimate for 61 Cygni, shortly after Wilhelm's paper was published. He, too, rejected position angles from the determination of parallax and he, too, increased his original estimate --but not so dramatically as Wilhelm had done.

In 1841 , the Royal Astronomical Society awarded Bessel its Gold Medal --the second he had received-- for his determination of the parallax of 61 Cygni. Although this recognition implied that Bessel's contemporaries regarded him as the first to achieve success, the work of Henderson and Wilhelm Struve was not ignored. Sir John Herschel, President of the Society that year, made the award and his remarks on that occasion are often quoted. They bear another repetition because they convey so well the excitement with which astronomers hailed this major achievement:

> I congratulate you and myself that we have lived to see the great and hitherto impassable barrier to our excursions into the sidereal universe; that barrier against which we have chafed so long and so vainly (aestuantes angusto limite mundi) almost simultaneously overleaped at three different points. It is the greatest and most glorious triumph which practical astronomy has ever witnessed. Perhaps I ought not to speak so strongly -- perhaps I should hold some reserve in favour of the bare possibility that it may all be an illusion - and that further researches, as they have repeatedly before, so may now fail to substantiate this noble result. But I confess myself unequal to such prudence under such excitement. Let us rather accept the joyful omens of the time, and trust that, as the barrier has begun to yield, it will speedily be effectually prostrated. Such results are among the fairest flowers of civilization.[23]

Because Wilhelm and Bessel both worked on and wrote about the determination of parallax over a considerable time, the question of which deserves the credit for being the first has not always seemed clear to their successors. The debate hinges on whether or not the value that Wilhelm presented for the parallax of Vega in the *Mensurae Micrometricae* is accepted as a genuine determination. Henderson, therefore, does not have a claim to priority since --although his measurements were made first-- both Wilhelm and Bessel published results before he did. A complication is that Wilhelm's 1837 value is closer to the modern one ($0\overset{''}{.}133$) than is his 1840 determination. With hindsight, his earlier value appears the better, despite its relatively large uncertainty. The Soviet historian of astronomy, Z.K. Sokolovskaya, has published part of a letter written by Wilhelm to Paul Fuss, Permanent Secretary of the Imperial Academy, which was read to the Academy on January 13, 1837 and is preserved in the archives of the U.S.S.R. Academy of Sciences:

> I trust I have been able to show that observations are now incomparably more exact, and parallax cannot escape us even if it consists of as little as 1/10th of a second of arc. It is possible to show from photometric data and also by calculations based on several double stars that a parallax of $0\overset{''}{.}1$ or a little larger exists to a high degree of probability. Since the general measurements of double stars have been completed, I began a special series of observations in this connection on the bright star α Lyrae, the most brilliant star in the

Sir John Herschel. (Reproduced with permission from the collection of presidential portraits of the Royal Astronomical Society.)

northern sky. Observations of its position with respect to a nearby 11th magnitude star about 43" distant and, I can show, not physically associated with it, are not yet complete enough to be absolutely decisive. Nonetheless calculations give a parallax of 0".14, so that, in all probability, the parallax lies between 0".10 and 0".18. If it turns out, as I hope, that further calculations confirm this result, this would constitute the important discovery that α Lyrae is at a distance from the solar system of 1 million solar distances.[24]

Except that it makes clear that Wilhelm had arrived at his first result very early in 1837, this letter does not add much to the *Mensurae Micrometricae*, which it resembles by its tentative tone.   How much more confident and forthright Wilhelm sounded in 1840! In his lifetime he did not claim to have measured the parallax of α Lyrae with certainty before he finished the 1837-8 observations, the results of which he did not submit for publication until well after the appearance of Bessel's paper.   In 1848, in the *Desription de l'Observatoire Centrale* , writing of the Pulkovo heliometer, Wilhelm said:

The heliometer, to the construction of which Fraunhofer had given himself, is without doubt one of the most perfect instruments of modern astronomy. It is by the admirable application that M. Bessel has made of it, that the science has been enriched by one of the greatest discoveries of our century, [that] of the knowledge of the annual parallax of the double star 61 Cygni.[25]

These are hardly the words of a man who felt himself cheated of the credit for a great discovery: they read, rather, like a generous concession of the priority to Bessel, by the only man who could have possibly contested it.

If Wilhelm had claimed his 1837 result as definitive, he would have been promptly criticized by his contemporaries on the grounds of its large probable error. They could not know, as we do, that the result is "right". If we ignore the large probable error, we have no way of ruling out even earlier claims, by others, that they had measured parallaxes that we now consider to be false.

Sokolovskaya also quotes a letter that Otto wrote to Lord Rosse in 1852 in which, at least by implication, Wilhelm's priority over Bessel is claimed. Otto's considered opinion, however, is probably reflected in the *Erinnerung*, where he writes that in the *Mensurae Micrometricae* his father

...finally made the first attempt crowned with success to determine the distance of a fixed star (α Lyrae) from our solar system, or at least to set narrow limits on it.[25]

Moreover, in the same year that Otto wrote the letter to Lord Rosse, he also published a Latin account of the measurements of the parallax of Vega (including his own result of 0".143) on the occasion of the silver jubilee of Dorpat University.[27] In this, he quoted --apparently with approval-- from John Herschel's address already cited, in which Herschel recognizes the importance of Wilhelm's early work, prior as it was to Bessel's, but still concludes that the latter's results were the ones that produced conviction.

It is beyond dispute that Wilhelm made real contributions to the determination of stellar parallax well before Bessel's successful work on 61

Cygni and that his arguments and observations, and especially the result published in 1837, helped to stimulate Bessel to *his* success. Both men had made systematic attempts, over long periods, to measure stellar parallax and both men were "distracted" by their many other scientific activities. It was largely a matter of luck which of them would succeed first, but contemporaries obviously believed that Bessel deserved the prize. Fortunately, he and Wilhelm were very good friends --too close to waste their time on arguments about priority-- and we hardly respect the memory of either man by raising such questions now. Let the last word remain with Wilhelm's great-grandson, the younger Otto Struve:

> It is of some historic interest that, among the memories transmitted to me by my family, the brightest was the high recognition accorded to Bessel by my great-grandfather, Wilhelm Struve. There was never any quarrel between these two astronomers, and they maintained the closest bonds of friendship until the death of Bessel in 1846. W. Struve never made any claims other than those I have quoted,* and I believe that some of the more recent claims on his behalf by others are exaggerated. It is difficult to find fault with those accounts that attribute to Bessel the first determination of a fully convincing stellar parallax.[28]

NOTES

1.   J. Jackson, 1922, Observatory, Vol. 45, pp. 341-352. Jackson's account has influenced much of my discussion. This particular result was first given by F.G.W. Struve in Dorpat Observations, Vol. 3, 1822 and was rediscussed by C.A.F. Peters in 1853.

2.   F.G.W. Struve, 1837, Mensurae Micrometricae. The portions relevant to this chapter have been translated by the Rev. Robert Main (Memoirs of the Royal Astronomical Society, Vol. 12, pp. 1-60, 1842) whose version is quoted here with permission of the Society. Page numbers refer to Main's translation. Wilhelm's criticism of his own observations is found on p. 20.

3.   F.G.W. Struve, 1837, ibid., p. 21

4.   F.G.W. Struve, 1837, ibid., p. 22

5.   F.G.W. Struve, 1837, ibid., p. 17

6.   F.G.W. Struve, 1837, ibid., p. 18

7.   F.G.W. Struve, 1837, ibid., pp. 18-19

8.   J. Ashbrook, 1954, Observatory, Vol. 74, p. 213 drew attention to this letter.

9.   A.H. Batten, B. Campbell and J.M. Fletcher, 1984, Publications of the Astronomical Society of the Pacific, Vol. 96, pp. 903-9.

10. F.G.W. Struve, 1837, Mensurae Micrometricae. This passage, translated by J.H. Walden, is quoted with permission from A Source Book in Astronomy (Harlow Shapley and Helen E. Howarth eds.), 1929, McGraw-Hill Book Co., New York, p. 213.

11. J. Jackson, 1922, see ref. 1. The tables of stars are given by A.S. Eddington in Stellar Movements and the Structure of the Universe, 1914, MacMillan and Co., London, Chap. 3.

12. F.G.W. Struve, 1837, Mensurae Micrometricae, see ref. 2, pp. 20-21.

13. F.G.W. Struve, 1837, ibid., p. 23.

14. Otto Struve, 1957, Sky and Telescope, Vol. 16, pp. 69-72 (p. 71).

15. F.W. Bessel, 1838, letter of October 23 to J.F.W. Herschel, translated and published by the latter in Monthly Notices of the Royal Astronomical Society, Vol. 4, pp. 152-161 (by whose permission it is quoted here) and reprinted in A Source Book in Astronomy (see ref. 10), pp. 216-220.

---

*   Essentially those also quoted here.

16 F W Bessel, 1838, ibid , pp 152-3

17 F W Bessel, 1838, ibid , p 153 See also F W Bessel, 1840, Astronomische Nachrichten, Vol 17, pp 257-276, (p 258)

18 F W Bessel, 1838, Astronomische Nachrichten, Vol 16, pp 65-96

19 T Henderson, 1839, presented his paper orally to the Royal Astronomical Society in January of that year Subsequent published discussions of the observations appeared in that Society's publications Monthly Notices, Vol 4, pp 168-9, 1839, Memoirs, Vol 11, pp 61-68, 1840 and Vol 12, pp 329-372, 1842

20 F G W Struve, 1840, Astronomische Nachrichten, Vol 17, pp 177-179, (p 177) This and subsequent extracts quoted with permission

21 F G W Struve, 1840, ibid , p 178

22 F Roberts, 1694, Philosophical Transactions of the Royal Society, Vol 18, pp 101-3

23 J F W Herschel, 1843, Monthly Notices of the Royal Astronomical Society, Vol 5, pp 89-98 (p 97), quoted with permission The Latin means seething at the very edge of the world

24 F G W Struve, 1837, letter of January 13 to Paul Fuss The letter quoted in Z K Sokolovskaya, Vestnik Akademii Nauk, U S S R , 1979, pp 132-6 The paper is known to me only through an unpublished translation by Michael Meo, from which he has kindly permitted me to quote

25 F G W Struve, 1845, Description, p 205

26 O W Struve, 1895, Erinnerung, p 38

27 O W Struve, 1852, Narratio de parallaxi stellae   Lyrae   Dorpat University

Second of Jensen's portraits of Wilhelm Struve, painted in 1843   This is copied from the original lithograph, now in the possession of Nils Lindhagen

CHAPTER 9

THE EARLY PULKOVO YEARS

Wilhelm's way of life changed in many ways when he moved to Pulkovo; one of the most striking of these changes was his withdrawal from active observation. "From 1843 on" Otto tells us "Father's personal participation in observing became rarer".[1] To be sure, in that very year he led a chronometer expedition to determine the longitude difference between Pulkovo and Altona, and he continued for some years to use "his" prime-vertical instrument, as Bond's and Airy's accounts show (Chapter 7), for the determination of the constants of nutation and aberration. Apart from this, however, he made very little use of the splendid equipment that he had provided for the Central Observatory. No doubt the administration of this new institute, with its large staff, absorbed much of his time and energy at first --although he soon began to hand that over to Otto, and at least did not have extra responsibilities like the supervision of the Cathedral Park or of the University Fire Service, that he had had at Dorpat. Undoubtedly, Wilhelm withdrew from observing partly in order to publish his earlier work. The *Mensurae Micrometricae* was published just before the move to Pulkovo, but the third volume of Wilhelm's great trilogy on double stars -the *Positiones mediae* of 1852- in which he gave accurate positions for all the double stars in the 1827 *Catalogus Novus*, occupied him soon afterwards. Then he set to work on the three-volume account of the arc measurement (Chapter 4), having produced the *Description* and *Etudes* d'astronomie stellaire* while he was still working on the *Positiones mediae*. It is almost as if Wilhelm sensed that his time was limited and that he must work hard to complete his major projects while he could. There may have been one other factor: Wilhelm was nearly fifty years old and perhaps his eyesight was no longer equal to the accurate measurement of double stars. If this were so, he might well have preferred to make no such observations at all, rather than to publish ones that would compromise his reputation. While there is no record that he complained of deteriorating eyesight, his son Otto did after *he* had turned fifty[2] --and there is a strong hereditary trait in these matters.

Wilhelm did not at first neglect outside affairs entirely. In the winter of 1841, he delivered a course of popular lectures on astronomy to the nobility of St Petersburg (the course that Grand Duke Konstantin attended --Chapter 7) which was repeated the following year for the people. Wilhelm's attractive *ex tempore* style of speaking drew large crowds, even though he spoke in German. A few years later, in 1845, Wilhelm joined with his former assistant Baron Wrangel, his former fellow-student K.E.von Baer and the Grand Duke himself in the foundation of the Russian Geographical Society.

---

*   I have followed the convention of omitting the accent on a capital letter.

DESCRIPTION

DE

L'OBSERVATOIRE ASTRONOMIQUE

CENTRAL

DE

POULKOVA,

PAR

F. G. W. STRUVE,

MEMBRE DE L'ACADÉMIE IMPÉRIALE DES SCIENCES DE S¹-PÉTERSBOURG, PREMIER ASTRONOME ET DIRECTEUR
DE L'OBSERVATOIRE CENTRAL.

St.-PÉTERSBOURG, 1845.

IMPRIMERIE DE L'ACADÉMIE IMPÉRIALE DES SCIENCES.

Title page of the "Description" from a copy in the possession of the Dominion Astrophysical Observatory.

In 1841 also, Wilhelm's old father Jacob died. They had last seen each other in 1838, and Wilhelm received the news in a letter from his sister, Christiane Henop. He, in turn, wrote to his nephew and foster-son in Dorpat:

> I was for a long time overcome by this report, which moved me to great sadness   His memory is a sacred one for us   We his children and grandchildren can and must bless it   We all owe him most, next to God, for what we are [3]

Two years later, Wilhelm visited his widowed mother during the course of his chronometer expedition to Altona. This expedition was undertaken to improve the knowledge of Pulkovo's precise geographical position within Europe (the occultation observations described in Chapter 7 were mainly concerned with the relative longitude of European and American observatories). While the position of Pulkovo relative to the old St Petersburg observatory had been well determined (Chapter 6) before any buildings were constructed, longitude differences between St Petersburg and, say, Paris or London were not well-known in 1839. A first step was to determine the difference between St Petersburg and Altona by transporting a large number of chronometers back and forth several times between the two points. At each terminus, astronomical observations were made to check the rate of each chronometer. The expedition in 1843 linked Moscow, St Petersburg, Warsaw and Altona, which last had been linked to other European observatories by the survey of the 1820s in which Wilhelm had taken part (Chapter 4). Between St Petersburg and Altona, the chronometers could be transported by sea from Kronstadt to Lübeck, with only short sections of land travel (more likely to affect the rate of the chronometers) at each end. Schumacher had measured the longitude difference between Altona and Greenwich in 1824, but Wilhelm decided to repeat the measurement in 1844. Tsar Nicholas, approached on Wilhelm's behalf by Uvarov, assigned 5,212 silver roubles to this stage of the project. Although Wilhelm had actively supervised the earlier expedition, he assigned that of 1844 to Otto and Döllen, who received some help from Hansen of Gotha. Nevertheless, Wilhelm travelled to Altona and to England, primarily to negotiate the assistance that they would need from Schumacher and Airy. It seems to have been this visit that cemented the friendship between Airy and both Struves. Wilhelm also visited South and John Herschel once again, and went to the university observatories at both Oxford and Cambridge. The former university awarded him its highest honour, the doctorate of civil law, and then demanded (after the ceremony) a fee of six guineas. Scandalized, Airy --who accompanied him-- refused payment on Wilhelm's behalf: the request was apparently withdrawn.[4]

Forty-four chronometers were transported eight times each way from Greenwich to Altona. Nearly twice as many had been used the year before between St Petersburg and Altona, but experience showed that it was better to select the most accurate clocks and to transport a smaller number. A regular steamer service plying between Hamburg and the Port of London was used. Indeed, at Hamburg, the ship used a dock in Altona, near Schumacher's observatory. In London, a special order was needed from the steamship company to authorize the captains to pick up and set down the chronometers just off Greenwich. Otto did not wish to go right up the Thames to the Port of London because he would have had either to transport the chronometers by land, or to transfer them to a river boat --which would have taken too long. He was also concerned about the loss of time that might be caused "by the strict laws of the English customs"[5], but the British government --perhaps prodded by Airy-- made special arrangements for him. Perhaps in return for this, the printed account published (in French) by the St Petersburg Academy, was dedicated to "Her Majesty Victoria Queen of Great Britain and of Ireland etc." Airy and Schumacher arranged for astronomical observations to be made at their observatories, and provided facilities for Otto and Döllen to make their own. Great precautions were taken to minimize personal errors; Otto and Döllen each spent some time at each terminus and they carefully compared their methods of reading the chronometers. The possibility of residual some little hopes, that a more complete acquaintance of our astronomical establishment, as given by the book, will engage You perhaps to a voyage to Russia.

The present year has been a very remarkable one for interesting astronomical discoveries, those of two new planets, a double comet, and a great number of other comets. The discovery of the Transuranian planet, now christened by the name of Neptun, in that way, in which it has been made, is the most eminent fact in the annals of the astronomical mens century, and places Le Verrier to the highest rank of scientific men. I dare say that this planet is to me particularly interesting, for I was search to for it, 10 years ago, as I made the perlustration of nearly +20000 stars from the northern pole to −15° of declination, for the construction of my newer catalogus stellarum duplicium. Now I know, by what reason I did not find the planet. In the years 1824 to 1826 it was in more than 20 degrees of southern declination beyond the limits of my perlustration. Micrometrical measures, made with our large telescope, have given an actual apparent diameter of 2,"7 to 2,"8, by agreeing observations made by myself and Otto. Such a diameter of a star would not have escaped to the excellent Dorpat telescope.

I beg You to present my compliments to Your ladies a to all benevolent English friends, who will remember especialy to captain Smyth and to the Rev. Dr Whewell at Cambridge. I think both will be not displeased to get a copy of the Description and of the Catalogus of our library.

The completion of our library continues to be an object of care and predilection for me. You will excuse, I hope, when I rise to recommend the interests of our library to Your particular care. I venture even to call on Your intervention for some particularities.

Extract from Wilhelm's letter of October 14/2, 1846, to Airy. (Reproduced by permission of the Royal Greenwich Observatory.)

magnetism in the hull of the ship affecting the chronometers had worried Wilhelm in 1843, and all chronometers used that year had been turned through 180° each day on board. Airy, however, had made experiments on magnetism in steamships, and advised Wilhelm in 1844 that this precaution was unnecessary. The first voyage from Altona began on June 11th and each return trip took about a week. The last voyage ended in Altona on August 8th. The difference in longitude between Greenwich and Pulkovo was found to be $2^h 1^m 18^s.674$, with an uncertainty of $0^s.057$ (within one-fifth of a second of the modern value).

In 1843, both Wilhelm and Otto formally adopted Russian nationality. Since 1834, when Wilhelm reached the civil-service rank of *wirklicher Staatsrat* (in Russian *deistvitel'nyi statskii sovetnik*), he attained hereditary nobility in Russia, so it may seem surprising that he did not formally change his citizenship for another nine years. Both he and Otto remained culturally German, but Otto remarks that Wilhelm's 35-year residence in Russia had estranged him from the German Fatherland, particularly as the political situation worsened there.[6] By this, Otto may simply have meant the increasing Prussian dominance in the move towards German union, or he may have been referring more specifically to the claims and counter-claims of Denmark and the German union to sovereignty over the Struves' native Schleswig-Holstein. Wilhelm must certainly have sensed the increasing tension between the German-speaking and Danish-speaking inhabitants when he visited Holstein in 1843. Five years later, in Europe's year of revolutions, the tension produced open rebellion, which cost Wilhelm's friend Schumacher much suffering and privation, and may have hastened his death in 1850.

Despite the labour of writing some of his own major publications, Wilhelm still kept up with developments in astronomy. On October 14/2 1846, he wrote to Airy:

> The present year has been a very remarkable one for interesting astronomical discoveries, those of the two new planets, a double comet, and a great number of other Comets. The discovery of the Transuranian planet, now christened by the name of Neptune, in that way [Wilhelm's emphasis], in which it has been made, is the most eminent fact in the annals of the astronomy in our century, and places Le Verrier to the highest rank of scientific men. I dare say that this planet is to me particularly interesting, for I was searching for it, 20 years ago, as I made the perlustration [i.e. the census of double stars] of nearly 120,000 stars from the northern pole to $-15°$ of declination, for the construction of my novus catalogus stellarum duplicium. Now I know, by what reason I did not find the planet. In the years 1824 to 1826 it was in more than 20 degrees of southern declination, beyond the limits of my perlustration. Micrometrical measures, made with our large telescope, have given an actual apparent diameter of $2''.7$ to $2''.8$, by agreeing observations made by myself and Otto. Such a diameter of a star would not have escaped to the excellent Dorpat telescope.[7]

Wilhelm wrote this letter in complete ignorance of the storm that, even then, was breaking around his friend's head over just this issue of the discovery of Neptune. Had Airy reacted more quickly to the work of the young J.C. Adams, credit for the discovery of the new planet might have gone entirely to English astronomers. Nationalistic hackles were raised by this thought, and Airy's role has been controversial ever since. As mentioned in Chapter 7, Wilhelm had to go to England in the summer of 1847. At Cambridge University, he attended the installation of Prince Albert as Chancellor and later, at the home of his friend Sir John Herschel, met both Adams and Le Verrier, the two men who had predicted the position of Neptune. Both of them impressed him, and he travelled back to France with the latter (*en route* to Altona), thus beginning a friendship that lasted until his own death, and which Otto continued until Le Verrier died. Although this famous French astronomer was not easy to get on with, he and Wilhelm appeared to have been bound by mutual respect, even though Wilhelm had resisted a French suggestion that the new planet bear Le Verrier's name.

Between 1848 and 1852, Wilhelm was much too absorbed in the preparation of the *Positiones mediae* to make any long journeys abroad. He did travel to Dorpat in December 1852, to take part in the University's celebration of its golden jubilee, where, naturally, he was welcomed as one of the most distinguished sons of the *Alma Mater*, having been associated with the University throughout almost its entire existence. Clearing his arrears of correspondence at the turn of the year 1848-9, he still found time to write to Airy about the new transit instrument in Greenwich and Lord Rosse's giant (six-foot aperture) reflecting telescope:

> You are now about very important changes in Your observatory, for Your new meridian circle, when successfully finished, will produce a new epoch in Greenwich meridian-observations, made then with the most powerful instrument of this kind in the world. I see in this circumstance an engagement for coming once more to England, where I have likewise to visit Lord Rosse's great telescope. I regret very much that I have not been your companion on Your visit to Ireland, last summer. You certainly have given to his Lordship many good directions; but one thing he seems to want, an intelligent astronomer for regular application of his instrument. Help him to find out such an individual, and You render an eminent service to science. For I believe that the hon. Lord ist [sic] not an astronomer himself, and to[o] much occupied, if he be, by political business and his new dignity as president of the Royal Society, which I believe engages him to a frequent residence in London.

Wilhelm did not get to see Airy's transit circle for several years, and never saw Lord Rosse's telescope. His last major break from desk-work, before his illness, was in 1853, and even this was partly a preparation for writing the two great volumes of *Arc du Méridien*, which occupied him for the rest of his working life. (The third volume of this work consisted of diagrams only.) The chief object of the 1853 journey, Wilhelm himself wrote in the Introduction to that work, was to discuss the publication with his Swedish collaborators. He also wished to visit his friend Argelander, who had moved to Bonn. Because of quarantine regulations then in force, Wilhelm went to Germany first. In Altona, he requested from C.A.F. Peters, who had succeeded Schumacher, the loan of one of the standards used in the Braak base-line measurement of the 1820s. Then, as explained in the Introduction to the *Arc du Méridien*

> ...I went to see my old friend, M. Argelander at Bonn, and his beautiful observatory that I still had not seen since it was finished. The visit to Bonn was the more interesting to me, since M. Argelander had written to me that His Majesty the King of Prussia, recalling a

FWA Argelander crayon drawing by A Hohneck 1852 in the possession of Universitäts-Sternwarte Bonn (Reproduced with permission)

promise graciously made on the occasion of his visit to Pulkovo in 1842, had charged him to make a journey to visit the Central Observatory of Russia   M Argelander was thus disposed to accompany me to Sweden and to go from there, through Finland, to Russia [9]

Argelander did come back to Pulkovo with Wilhelm, but first they both visited Sweden and the Swedish geodesists   D Georg Lindhagen, soon to become Wilhelm's son-in-law, travelled there from Pulkovo to meet them   A

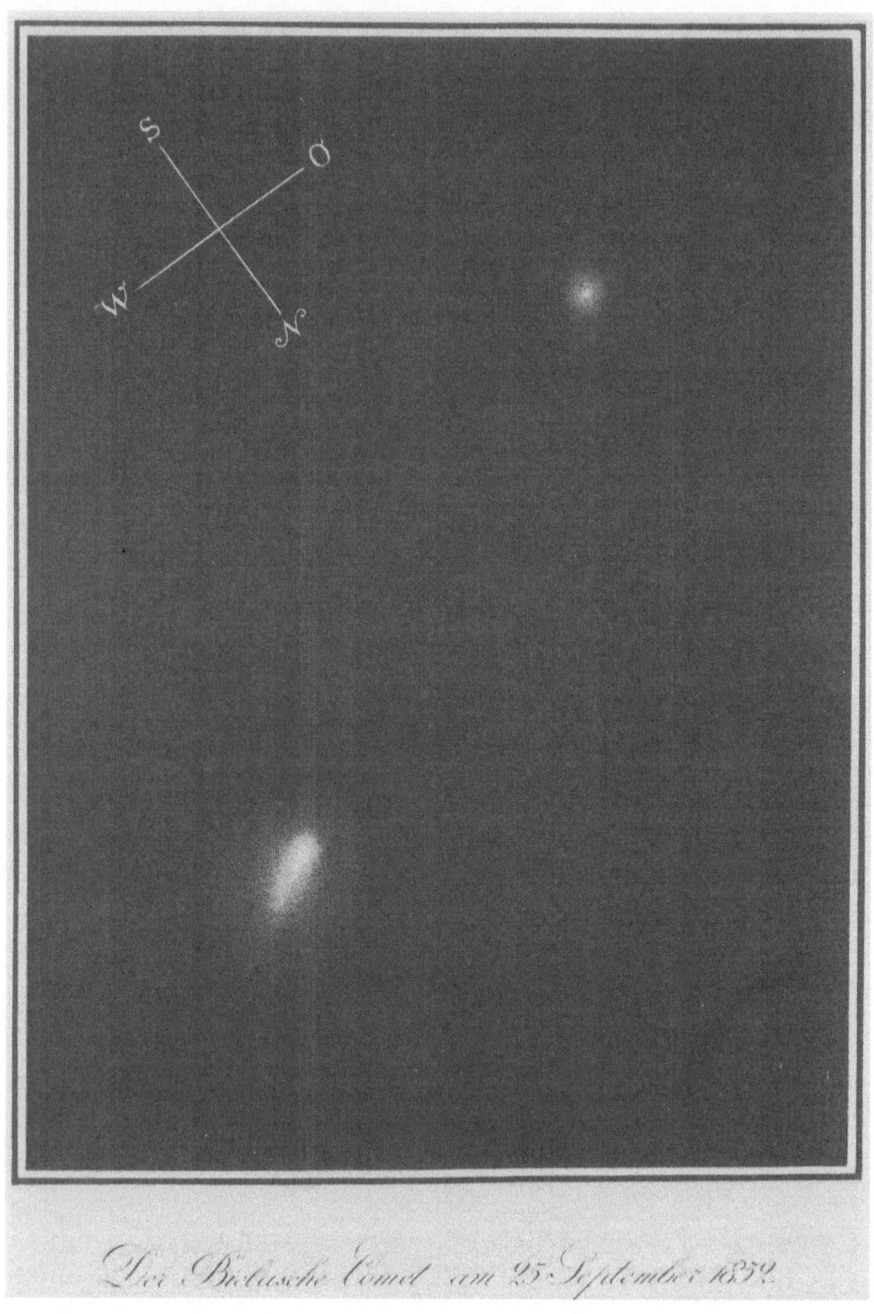

Otto's drawing of the two parts of Comet Biela as observed on 25 September 1852 -
-probably the last record of the dying comet. (From Mémoires de l'Académie
Impériale des sciences St-Pétersbourg.)

letter written by Wilhelm, from Bonn, to Lindhagen, sets July 14th as the date for them all to meet there.[10]   After Argelander left Pulkovo, Wilhelm settled down to work on the *Arc du Méridien*.

As Wilhelm began to restrict his activities in the 1840s, directing his energies into more specific channels, Otto perhaps reached the high point in his own career of scientific research.   From 1847, when he became his father's official deputy, he was forced more and more to be an administrator of research policy, rather than a researcher himself.   When Pulkovo was opened, Wilhelm had entrusted his son with the great new 15-inch refractor, with which Otto acquired a special skill (Chapter 7).   His first task was to conduct a survey for any double stars that his father might have missed with the Fraunhofer refractor.   He discovered over 500 pairs, still usually known by the serial numbers in his catalogue prefixed by the letters OΣ.   Most of the pairs were closer than the majority found with the Fraunhofer instrument, and there are several systems of interest among them.   The catalogue appeared with Wilhelm as the author, but the majority of the discoveries were undoubtedly Otto's.[11]   Contemporaries believed that Otto and his father had, between them, discovered all the double stars that could be found with such instruments. and the work of these two tended to inhibit further surveys.   Not until the end of the nineteenth century did Burnham, Aitken and Hussey reveal the rich opportunities for discovery that still remained.

Otto also used the large refractor for studies of the solar system.   He began the observation of the satellites of planets that his son Hermann and grandson Georg were to make into their own special niche in astronomy.   In 1902, Hermann received the Struve family's third Gold Medal from the Royal Astronomical Society just for his studies of the satellites of Saturn.   In March 1849, Otto wrote to Airy about his own satellite observations:

> I have continued to observe the satellites of Uranus and that of Neptune so regularly as I could do. Of the two principal satellites of Uranus I have now got a sufficient number of good measures to try a new, and I hope, exact deduction of the elements of their motion and of the mass of the planet. The third satellite has been seen several times by me and some good measures of it are taken. Nevertheless the number of these measures is yet to[o] small to furnish exact values of its elements. In the last autumn I have seen it twice or three times in a direction opposite to that in which I saw it constantly the year before.[12]

The English observer Lassell is usually credited with the discovery of the third and fourth satellites of Uranus in 1851, and it is still uncertain what Otto saw and measured in 1847-9.[13]   Otto also observed several comets with the telescope, particularly Biela's comet which used to travel around the Sun in a period of just over six years.   In 1846, it was found to have split in two --the first reliably recorded such event, since reports by classical authors had not previously been believed-- and was the "double comet" to which Wilhelm referred in the letter quoted above.   In 1852, the comets returned on schedule, but were never seen again with certainty after they left the vicinity of the Sun. At this last apparition, the comets were first seen from southern observatories, and they moved northwards as they receded from the Sun.   Otto, with his large telescope at a northern observatory, may well have been the last person ever to see them.   His observations are summarized in a memoir published by the St

Petersburg Academy,[14] illustrated with his own drawings. The drawings attest to Otto's skill as a draughtsman, just as Wilhelm's publication, nearly 20 years earlier, of a monograph on the 1835 apparition of Halley's comet had attested to *his*.

Contemporaries, however, considered Otto's determination of the constant of precession and the direction of solar motion to be one of his most important scientific studies. He began the work at Dorpat, as part of an observatory project to determine the fundamental constants of astronomy. Otto's share of the work was published as another memoir of the Academy in 1841.[15] Not until 1850, however, did the Royal Astronomical Society --the President of which was then Airy-- award the Struve family its second Gold Medal to Otto, for this work. Airy delivered the customary address at some length --the printed version extends to some twelve pages-- on February 8th of that year. After the formal announcement, Airy said:

> ...it is now my most agreeable duty to point out to you the state of the questions to which this paper relates, the need for a new investigation, the skill and care with which M. O. Struve's investigations have been conducted, and (in a word) the claims which the selection of the subject and the mode of treating it have presented for the honourable notice of the Society.[16]

Airy described at some length how precession, known since classical times (see Chapter 8), remained unexplained until Newton showed that the gravitational action of the Sun and Moon on the equatorial bulge of the Earth makes the Earth's axis describe a giant cone in space, once every 26,000 years. Airy remarked

> ...I have no hesitation in expressing my belief that, if I had lived in that critical time of science, the explanation to which I should have appealed as establishing the truth of the theory of gravitation would have been the explanation of precession.[17]

Before Otto's work, Airy tells us, the most generally accepted determination of the rate of precession had been made by that other family friend of the Struves, Friedrich Wilhelm Bessel. As a result of precession, the position that the Sun occupies at the spring equinox changes from year to year and, because this is the zero-point to which all stellar positions (right ascensions and declination) are referred, they, too are continuously changing. The determination of the rate of this change is made more difficult by the proper motions (transverse motions across the sky) of individual stars. Proper motions of many nearby stars are an appreciable fraction of the annual rate of precession; moreover, as Wilhelm Herschel showed, there is a systematic effect in proper motions that is a reflection of the Sun's own motion through space. The problem is to sort out the star's own motions, the Sun's motion and the effect of precession. Large numbers of stars have to be studied so that their individual motions, if they are randomly distributed in size and direction, will be averaged and have no systematic effect on the final result. Bessel had tried this in 1815 and 1819, but did not confirm the solar motion postulated by Herschel, which therefore remained controversial until yet another family friend of the Struves, Friedrich August Argelander, put its existence beyond doubt. Even Argelander's work, however, left the amount of solar motion very uncertain. Otto hoped to improve the determination of both solar motion and precession. He began by assuming that Argelander had determined the direction of solar motion correctly, then he selected 400 stars with accurately determined modern positions. (Well over half of these were double stars observed from Dorpat by Wilhelm.) Otto compared these modern positions with the positions given for the same stars, for the year 1755, in Bradley's catalogue (as edited by Bessel).

Airy's account of Otto's work was not completely uncritical. He pointed out which assumptions could be questioned and emphasized the difficulty of satisfactorily separating the effects of solar motion and precession. Otto himself had demonstrated that even a very small systematic error in Bradley's determinations of position (of a size that Airy did not believe could be ruled out) could have a large effect on the direction found for the solar motion,

...and [Airy was] therefore induced, by M. O. Struve's remarkable conclusion, to look upon all these determinations of the pole of solar movement with doubt.[18]

Despite the doubt, Airy thought highly of Otto's work and went on:

I have no hesitation in expressing my own opinion that M. Otto Struve has grappled with the various difficulties that present themselves, metaphysical and mathematical, better than any other astronomer whose work on the same subjects I have examined. The two investigations which relate to the determination of the direction and magnitude of the solar movement are, in my opinion, very admirable; but the third, which exhibits the amount of uncertainty in the result depending on venial or probable errors of observation, is, in my judgement, even more valuable. I esteem a rational doubter in science as I esteem a Niebuhr in history --as the only person who is likely to supply at last a firm foundation upon which a solid superstructure may be built.[19]

In these words Airy revealed something of his own character and accurately portrayed Otto, who was a painstaking critic --perhaps even to the point of inhibiting his own creativity. Airy suggested that Otto's work would not be the last word on these topics --he could not have foreseen that Otto's son Ludwig would be one of those who would try to improve on it-- and concluded:

M. O. Struve has given to the world a paper of the most important character, worthy of himself and worthy of the name which he bears.[20]

Otto could not go to London to receive his medal and, indeed, seems not to have known of the award until the printed address reached him in Pulkovo in April 1850. He promptly wrote to Airy:

This morning I have got the number of the Monthly Notices which contains your address on the award of the Astronomical Society's Medal to my inquiries on the Precession and Proper motion of the Solar Systems. Allow me now to express to you, as the President of that Society, my warmest thanks for the extraordinary and quite unexpected honour, which the Council has conferred on me, but still more to you, as the eminent astronomer, for the kind and benevolent words with which you have accompanied the award. For this moment I have several reasons to appreciate particularly the award of the Medal, but if anything could raise its value in my eyes, it was that I have got it under your presidency.

Pulkova 14 April 1850

My dear Sir

This morning I have got the number of the Monthly Notices which contains your address on the award of the Astronomical Society's Medal to my inquiries on the Precession and Proper Motion of the Solar System. Allow me now to express to you, as the President of that Society, my warmest thanks for the extraordinary and quite unexpected honour, which the Council has conferred on me, but still more to you, as the eminent astronomer, for the kind and benevolent words with which you have accompanied the award. In this moment I have several reasons to appreciate particularly the award of the Medal, but if anything could raise its value in my eyes, it was that I have got it under your presidency. Your friendly words have done well my heart and not less that of my father. I thank you most sincerely for them!

I have to answer now on two of your letters that of Dec 27 and that of March 23. All publications of the Academy which lately have appeared shall be delivered to us and will be sent to you by one of the first steamers. Also the Bulletin Physico-Mathematique will go the same way. One of these days my father will go to Lord Bloomfield to ask him if he will allow us to forward to you the singular numbers of that journal by a diplomatic messenger.

Letter from Otto to Airy, dated April 14, 1850. (Reproduced by permission of the Royal Greenwich Observatory.)

Your friendly words have done well my heart and <u>not less that of my Father</u> [Otto's emphasis.] I thank you most sincerely for them![21]

NOTES

1. O.W. Struve, 1895, <u>Erinnerung</u>, p. 54.

2. O.W. Struve, 1876, letter of February 1 to Simon Newcomb. (Library of Congress,, Simon Newcomb Papers, Container 40).

3. Nikolai Struve, 1915, <u>Deutsche Monatschrift fur Russland</u>, Vol. 57. I know this article only at second hand: this brief extract was provided by Nils Lindhagen.

4. Wilfred Airy (ed.), 1896, <u>Autobiography of Sir George Biddell Airy, K.C.B.</u>,(see Chap. 7, ref. 22), pp. 165-6.

5. F.G.W. Struve and O. Struve, 1846, <u>Expédition Chronométrique executée par ordre de Sa Majesté l'Empereur Nicolas Ier entre Altona et Greenwich pour la détermination de la longitude géographique de l'Observatoire Central de Russie</u>. St. Pétersbourg, p. 3. The whole account of the expedition, not just the quotation, is taken from this source.

6. O.W. Struve, 1895, <u>Erinnerung</u>, pp. 54-5.

7. F.G.W. Struve, 1846, letter of October 14/2 to G.B. Airy. (RGO 939 ff. 542-3).

8. F.G.W. Struve, 1848, letter of December 31/19 to G.B. Airy. (RGO 940 ff. 168-9).

9. F.G.W. Struve, 1860, <u>Arc du Méridien</u>, Vol. 1, p. XXXVIII.

10. F.G.W. Struve, 1853, letter of June 25 to D.G. Lindhagen. The original is in the writer's possession.

11. F.G.W. Struve, 1843, <u>Catalogue de 514 Etoiles Doubles et Multiples découvertes sur l'hémisphère céleste boréal par la grande lunette de l'Observatoire Central de Poulkova, et Catalogue de 256 Etoiles Doubles Principales ou la distance des composantes est de 32 secondes à 2 minutes et qui se trouvent sur l'hémisphère boréal</u>. Académie Impériale des Sciences, St. Pétersbourg. (Revised and corrected edition, 1850).

12. O.W. Struve, 1849, letter of March 27 to G.B. Airy. (RGO 941 ff. 209-210).

13. P. Mozel, 1986, <u>Journal of the Royal Astronomical Society of Canada</u>, Vol. 80, pp. 344-350, describes events leading up to the discovery of the third and fourth satellites of Uranus.

14. O.W. Struve, 1854, <u>Beobachtungen des Bielaschen Cometen im Jahre 1852 angestellt am grossen Refractor der Pulkowaer Sternwarte</u>, Memoires de l'Académie Impériale des Sciences de Saint Pétersbourg. 6ème série. Tome VI, pp. 1-26.

15. O.W. Struve, 1841, <u>Bestimmung der Constante der Praecession mit Berucksichtigung der eigenen Bewegung des Sonnensystems</u>. Mémoires de l'Académie Impériale des sciences de Saint Pétersbourg. 6eme série. Tome III.

16. G.B. Airy, 1850, <u>Memoirs of the Royal Astronomical Society</u>, Vol. 19, pp. 271-283, (p. 271).

17. G.B. Airy, 1850, <u>ibid</u>., p. 273. (Quotations by permission of the R.A.S.).

18. G. B. Airy, 1850, <u>ibid</u>., p. 281.

19. G.B. Airy, 1850, <u>ibid</u>., p. 281.

20. G.B. Airy, 1850, <u>ibid</u>., p. 282.

21. O.W. Struve, 1850, letter of April 14 to G.B. Airy. (RGO 942 ff. 396-7).

ÉTUDES
D'ASTRONOMIE STELLAIRE.

SUR

LA VOIE LACTÉE

ET SUR

LA DISTANCE DES ÉTOILES FIXES.

RAPPORT

FAIT

A SON EXCELLENCE

M. le Comte Ouvaroff,

MINISTRE DE L'INSTRUCTION PUBLIQUE ET PRÉSIDENT DE L'ACADÉMIE
IMPÉRIALE DES SCIENCES,

PAR

F. G. W. Struve,

DIRECTEUR DE L'OBSERVATOIRE CENTRAL DE RUSSIE ET MEMBRE
DE L'ACADÉMIE.

ST.-PÉTERSBOURG,
IMPRIMERIE DE L'ACADÉMIE IMPÉRIALE DES
1847.
(Für die Astronomischen Nachrichten besonders

Title page of the "Etudes". (Reproduced by permission of Swarthmore College Library.)

CHAPTER 10

"ETUDES D'ASTRONOMIE STELLAIRE"

Most of Wilhelm's major works have now been set in their context. There were many other papers --some quite short-- in *Astronomische Nachrichten*, the *Monthly Notices of the Royal Astronomical Society* and the various publications of the St Petersburg Academy, not to mention the early volumes of *Dorpat Observations*. There remains the book *Etudes d'astronomie stellaire*, published in 1847, which was important in its time and reveals a different side of Wilhelm's scientific personality.

The book is dedicated to "M. le Comte Ouvaroff, Ministre de l'Instruction Publique et Président de l'Académie Impériale des Sciences". In both capacities, Uvarov was Wilhelm's superior, so the book is in part a report to Wilhelm's political masters on the work of Pulkovo Observatory. Bearing the *imprimatur* of the Academy, signed by Paul Fuss, the book seems also to have been intended for a wider readership. The new results in it were certainly intended for Wilhelm's professional colleagues, but the book also has something of the character of an advanced popular work on astronomy. The term "stellar astronomy" meant, in particular, the study of the distribution of stars in space --a study being made possible by the first determinations of parallax.

The book opens with a strong historical introduction, praised by friendly and hostile critics alike. Wilhelm had made good use of the Pulkovo library and showed familiarity with some books that, even when he wrote, had been forgotten by all but historians. His introduction is still a useful account of the development of ideas of the Galaxy to which our Sun belongs. In this field, as in the study of double stars, Wilhelm gladly and generously recognized the pioneering work of "the immortal astronomer" William Herschel --whom he describes as "England's greatest astronomer". In both fields, Wilhelm aspired to be Herschel's disciple --but clearly recognized him as the master. The development of Herschel's views is described in over twenty pages, in which Wilhelm made good use of the annotated set of Herschel's papers with which he had been presented (Chapter 5). Of particular interest to Wilhelm were two papers Herschel wrote in 1784 and 1785[1] on "the construction of the heavens" and several subsequent papers, including the final one of 1818[2], in which the immortal astronomer recorded his gradually evolving ideas on the same topic. Wilhelm emphasized very strongly the changes in Herschel's ideas in those three decades.

In the early papers, Herschel drew attention to the marked concentration of all but the brightest stars to the plane of the Milky Way, and deduced that the stars belonged to a highly flattened system of a thickness small compared to its diameter. Herschel thought the Sun was near the central plane of this

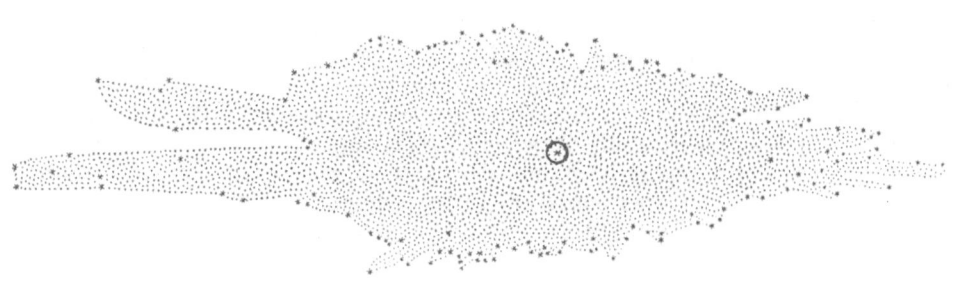

William Herschel's diagram of the Galaxy reproduced from his 1785 paper in Philosophical Transactions of the Royal Society. *The diagram shows a cross-section perpendicular to the plane of the Galaxy and the circle represents the sphere within which the naked-eye stars are found. (Reproduced by permission of the Royal Society.)*

system, not far from the point where the Milky Way divides into two halves -- as it does from Cygnus to Scorpio. He counted the number of stars visible in the field of his 20-foot telescope in about 2,400 different directions; he believed that these numbers were measures of the distances to which the stars extended, and that his telescope could penetrate to the limits of the system. He used his counts, or "gages", to construct a diagram showing the shape of the Milky Way system --a diagram still reproduced as "Herschel's system" in many modern texts.

By 1818, Wilhelm insisted, Herschel had abandoned this system According to Wilhelm, two things in particular had led Herschel to change his views, namely, his own discovery of the existence of genuine binary systems and the construction of his 40-foot telescope. The discovery that stars of very different apparent brightness are revolving around a common centre and must be at the same distance from us destroyed the assumption --essential to Herschel's early work-- that all stars have much the same luminosity. At the same time, Herschel came to realize that not all nebulae were resolvable into stars and, therefore, some of the luminous matter in the universe is not in the form of stars. With the large telescope, he made some more gages and realized, according to Wilhelm although Airy disagreed, that even this giant could not fathom the Milky Way itself. Thus, the number of stars seen in a given direction was not simply related to the extent of the system.

Wilhelm then summarized the work of his contemporaries, discussing variable stars, the measurement of parallax (which eluded Herschel, Wilhelm says, because his micrometers were too imperfect) and the motion of the Sun with respect to the stellar system. He mentions --with some doubts as to their reality-- the periodic changes in the proper motions of Sirius and Procyon which Bessel had discovered and (correctly) ascribed to companions, invisible at that time, of each of these stars. He even found space to mention the theory of Mädler --his successor at Dorpat-- that the entire Galaxy rotates about the brightest star in the Pleiades. The theory was never fully accepted, but Wilhelm pointed out that, as early as 1833, John Herschel had suggested (also correctly) that globular clusters would be stable if they rotate. Wilhelm thought it very bold to suppose the whole Galaxy to be rotating, and that such a supposition "is a little too dangerous in the present state of science, and that it is open to very serious objections, both theoretical and from the celestial phenomena".[3] Galactic rotation, of course, has now been clearly demonstrated, but the *Centralsonne* --if there be such a unique body-- is nowhere near the Pleiades. Wilhelm emphasized the progress that had been made since the time of the elder Herschel in the study of double stars --several orbits having been determined-- but pointed out that only John Herschel had progressed beyond his father in the study of the nebulae. As for "the two clouds named after Magellan", Wilhelm inclined to the opinion of Lacaille that they are "detached portions of the Milky Way".[4] No advance had been made since William Herschel's time in the knowledge of the Milky Way and Wilhelm emphasized again his puzzlement that astronomers insisted on using Herschel's 1785 model, although it had been abandoned by its author. (Herschel did not explicitly withdraw his 1785 model in his later papers, but Wilhelm does seem to be correct in saying that Herschel gave up the assumptions on which the model was based.)

The *Etudes* contains just over 100 not very closely printed pages and about 60 pages of notes, and these introductory sections take up more than half the pages of the main text. In the remainder Wilhelm develops his own ideas, beginning with a summary of the Introduction he had written in the previous year for the star catalogue published by the Cracow astronomer Maximilian Weisse.[5] This catalogue was based on observations by Bessel and Argelander, made from Königsberg, of stars between declinations -15° and +15°, and it is basic to the rest of the *Etudes*. It contains 31,895 stars, supposedly down to the ninth magnitude: 807, however, were outside the declination limits and three were of tenth magnitude --leaving 31,085. Just as in the *Catalogus Novus* (Chapter 5) Wilhelm allowed for the inevitable incompleteness of catalogues. Since most of the stars in this one had been observed only once, he believed completeness corrections to be very important. By comparison with overlapping catalogues, Wilhelm estimated that Weisse's contained about 82 per cent of the stars down to the adopted brightness limit and within its zone of declination. Many more fainter stars than brighter ones were missed, however, and Wilhelm estimated that the catalogue contained only about 77 per cent of the eighth-magnitude stars and 52 per cent of the ninth-magnitude ones, and that the zone béing studied contained about 52,000 stars down to the adopted magnitude limit. Dividing the zone by hours of right ascension, he showed that the stars displayed the same concentration toward the Milky Way that he had found in the *Catalogus Novus*. Wilhelm concluded that the Sun was near the centre of the disk and surrounded on all sides by stars which, however, are not distributed uniformly. He correctly identified the maxima and minima of the distribution, and claimed that the density, even of stars brighter than ninth magnitude, falls off on either side of the Milky Way, successive bands parallel to that feature containing fewer stars than the band before them. He believed that the Sun was not exactly in what we should call the plane of the Milky Way, but a little to one side --by about one-sixth of the mean distance of stars of the sixth magnitude. He was, of course, well aware of the fluctuations in density of the stars, even in the Milky Way itself. He stressed again that Herschel could not fathom the Milky Way with his 40-foot telescope and that, therefore, we could say nothing about the exact dimensions of the system.

Wilhelm then introduced the more controversial part of his book with these striking remarks:

> The composition of this report had furnished me with an opportunity to return to the study of the Milky Way. This phenomenon is so puzzling, at first glance, that one is almost tempted to give up a satisfactory explanation. However, the man of science ought not to recoil, either before the obscurity of a phenomenon or before the difficulty of a research. Whether he is in possession of earlier work, or whether he tries to increase the knowledge of a phenomenon by new, precise observations; he can be sure of a certain success in his studies, if he employs calm speculation, without abandoning himself to an excited and preoccupied imagination. However little ground he gains, he will always enlarge it by returning to his problem with that persistence which is the indispensable condition of study. It is thus, guided by analysis and calculation, that he can even derive unexpected results which, however, enjoy considerable certainty.[6]

Not all Wilhelm's contemporaries agreed that his speculation was calm. While he maintained that we could not know the true extent of the Milky Way, he argued that we could study the law of its condensation towards the central plane. If we knew that law, we could deduce the number of stars to be seen in a given telescope, in whatever direction it was pointed. Conversely, from star counts we should be able to derive the law --which is precisely what he tried to do in pp. 67-82 of the *Etudes*. Although Wilhelm believed that the Sun did not lie in the central plane of the Milky Way, the displacement he supposed it to have --about $1/250^{th}$ part of the radius of a sphere containing all stars visible in Herschel's 20-foot telescope-- was too small compared with the dimensions of the Milky Way itself to affect the analysis. Wilhelm showed that the density of stars falls off very quickly on either side of the Milky Way. As well as Weisse's catalogue, Wilhelm used selected star "gages" of Herschel's and approximated the system of stars by a series of thin parallel layers on either side of the Milky Way, each successive layer containing fewer stars. Thus he showed that, looking out from the Sun at an angle of 60° to the (approximate) plane of the Milky Way, one looked through a layer in which the density of stars was only 1/200 of that in the Milky Way itself. He tried to express all distances in terms of some unit --such as the mean distance of stars of the first magnitude-- and thus built up a picture of the visible part of the system as highly flattened and of dimensions that could be determined once the basic unit was known.

In the next section of the *Etudes*, beginning on p. 83, Wilhelm introduced the idea that starlight suffers absorption on its journey through space. He was not the first to propose this and quoted from a memoir on the transparency of space by Olbers[7] in which the paradox of the darkness of the night sky --now known by Olbers' name-- was introduced. Long forgotten, the paradox was much discussed in the years when the steady-state cosmology was popular. If stars are distributed uniformly through infinite space, the sky should be everywhere as bright as the Sun. Olbers thought that either the universe could not be infinite (or, at least, could not contain an infinite number of stars) or some of the light emitted by the stars must be lost *en route*. He calculated that starlight need lose only 1/800th part of its intensity in travelling a distance equal to that from Sirius to the Sun in order to "save the phenomena". Wilhelm gave Olbers full credit for reaching this conclusion before him, and even pointed out that de Chéseaux had anticipated Olbers in 1744, although the latter apparently overlooked the fact.

Wilhelm attempted a more accurate calculation of the amount of absorption needed to avoid Olbers' paradox. He began with Herschel's own estimate that the 20-foot telescope could reach about 61 times farther than the naked eye. This estimate, however, was based on the assumption that the pupil of the eye is exactly one-fifth of an inch in diameter, whereas light-grasp and visual acuity differ very much from one person to another. Wilhelm made an achromatic telescope of just over 0.2 inches in diameter, which admitted the same amount of light as Herschel's "standard eye", but formed clearer images, whatever defects the observer's own eyes might possess. With this, Wilhelm could see nearly twice (actually 1.83 times) as many stars as were listed in Argelander's catalogue of naked-eye stars. He deduced that Herschel's telescope should penetrate $\sqrt[3]{1.83}$ x 61.18 = 74.83 times as far as the average distance of the sixth-magnitude stars. Wilhelm believed that these stars were (on average) 8.87 times as far away as the first-magnitude stars. He also believed, however, that the relative numbers of stars of different magnitudes observed by Herschel showed that the 20-foot telescope reached only 25.7 times as far as the average distance of sixth-magnitude stars. Thus, observations showed the penetrating power of the telescope to be about one third that calculated from optical theory, and this Wilhelm ascribed to interstellar absorption. He calculated that we see only 1/25th of the stars that we should see if space were transparent and that this indicated that the light of sixth-magnitude stars had been diminished, on average, by about 8 per cent (or that of first-magnitude stars by about one per cent), on its journey to the Earth. Nearly a century later (in 1930) R.J. Trumpler finally convinced astronomers of the reality of interstellar absorption of starlight.[8] If his estimate of the amount of absorption is expressed in Wilhelm's terms, and Wilhelm's assumed average distance of first-magnitude stars is adopted, Trumpler's estimate of interstellar absorption is less than twice the above value. This is largely coincidence, however, since Wilhelm's attempt to estimate what are now called secular parallaxes --estimates of the distances of stars based on the part of their proper motions that reflect the Sun's motion -- led him to underestimate the distances of different classes of stars.

Wilhelm's hypothesis of interstellar absorption was not immediately accepted. He was ahead of his time, but he was right for the wrong reasons. Absorption of light provides no escape from Olbers' paradox, since the matter absorbing the light (and heat) must eventually begin to radiate like stars itself. Wilhelm can be forgiven for not realizing this. Thermodynamics was in its infancy (one of Helmholtz' basic papers was published in the same year as the *Etudes*) and astronomers did not then ascribe the great age to the universe that their modern successors do, although John Herschel seems to have understood the force of the thermodynamic objection. Note that Wilhelm's attempt to estimate the *amount* of the absorption does not depend on his belief that it was required by Olbers' paradox.

The *Etudes* concludes with a summary of a memoir (then still in press) in which C.A.F. Peters summarized the then current knowledge of stellar parallaxes. Peters considered 35 stars to have reliably determined parallaxes , but two of them --61 Cygni and Groombridge 1830-- with very large proper motions were excluded from the statistics. Rather riskily, Peters tried to determine mean parallaxes for the various magnitude classes from the remaining 33 stars. In particular, he found a mean parallax of 0".116 for second-magnitude stars. Wilhelm used this value to turn his relative distances into absolute ones, and calculated that Herschel's 20-foot telescope could penetrate just over 3,500 light-years from the Earth. Finally, Wilhelm used Otto's determination of the solar motion (in angular measure) to estimate the Sun's linear speed through space as 1.6 times the radius of the Earth's orbit in the course of a year. This figure became the basis of his estimates of secular parallaxes. The precise value found for the speed of the Sun depends on the group of stars observed, but modern values are about double Wilhelm's and his secular parallaxes are about half what they should be.

The *Etudes* had a mixed reaction, but it certainly was not ignored. Two British reviewers, Airy and Sir David Brewster, praised it highly. Brewster, a physicist remembered mainly for his studies of the polarization of light, was at that time Principal of United College in the Scottish University of St Andrews. Particularly interested in various forms of the teleological argument, he was a prolific writer of popular science articles and book reviews. So prolific was he, that it is difficult to track down all his reviews, but the tone of this one can be guessed from Wilhelm's letter (in English) of December 1848 to J. Lee of Hartwell House:

> I have to request that you will accept my late though not less sincere thanks for your kind letter of March 20, and the different printed papers added. The articles of Sir D. Brewster and of the Astronomer Royal on my Etudes d'astronomie stellaire have given me the greatest satisfaction, and I dare say I am proud of the honour done to this little work by those eminent philosophers.[9]

Airy's extensive comments, on the other hand, form part of his presidential address to the Royal Astronomical Society for 1847.[10]

> A remarkable work on the Distribution of the Stars has been published by M. Struve, under the title of Etudes d'astronomie stellaire. The treatise bearing this name contains an epitome of the whole of the author's views upon this subject...[11]

Airy gave a full and clear summary of the contents of the book and closed with this comment:

> We have considered that the classical character, the ingenuity of the mathematical processes (in small details as well as in the more ostensible steps), and the importance of the results, of this remarkable treatise, deserved more than usual attention. The general character of the conclusions, we think, can scarcely be doubted, although their application

must be subject to great irregularities. We conceive, for instance, that they exclude the possibility of such an annular arrangement of stars as has been depicted in some works intended to explain M. Mädler's views detailed in the <u>Centralsonne</u>. If we might be permitted to comment on it in the way of criticism, we should remark, that the repeated assertion of the ascertained unfathomability of the heavens by the 40-foot telescope is scarcely supported by the expression of Sir W. Herschel, in the only place (we believe) where it is mentioned as probable; but the bearing of this assertion on the results of this treatise is not important.[12]

Other distinguished colleagues and friends reacted less favourably. Encke, who had worked with Wilhelm on the Braak base-line in 1820, helped with the acceptance tests of the Pulkovo instruments, and was by then Director of the Berlin Observatory, wrote a review in the *Astronomische Nachrichten*, longer than Airy's and as hostile as the latter's had been friendly. With the merest nod in the direction of courtesy, Encke plunged *in medias res*:

Staatsrath Struve has had the goodness to send his Etudes d'astronomie stellaire to the <u>Astronomische Nachrichten</u>, and thus has brought it to the knowledge of all experts in, and lovers of astronomy. Its most inviting content concerns questions that everybody certainly has often put to themselves. Everyone, therefore, will have the more cause to make themselves acquainted with it. The very great and everywhere-recognized, firmly based authority of a Struve will not fail to secure immediate entry for the numerical data, which are the result of this investigation, into all writing on this subject, particularly the so-called popular [ones]. Just on this account, however, I hope that an excuse will be found if I try, partly to establish more closely the dependence of the results on the assumptions, partly also not to hesitate to express my doubts about both, openly and without reserve. If, with lesser esteemed astronomers it is unnecessary to explain and even to challenge opinions, which by their nature cannot rest on strict principles, and [which] may, therefore, be left to the reader's own judgment, so the expressly presented assumptions of a Struve appear to need careful examination, and the discussion of them can only be beneficial to science.[13]

Even on his first quick reading, Encke wrote, he had doubts about the table of the relative distances of stars in the various magnitude classes, which Wilhelm had claimed to be based entirely on observation --without any hypothesis having been adopted. Encke denied this claim and suggested that by the phrase "based only on observation without any hypothesis" Wilhelm meant "no hypothesis for which he cannot have more or less reason to derive from the observations".[14] Encke identified five hypotheses that he believed Wilhelm to have made:

(i)   the distribution of stars is not uniform over the whole sky, but is roughly so in the Milky Way itself;

(ii)   the apparent brightness of a star is a rough measure of its distance;

(iii)   within the Milky Way, stars are distributed roughly uniformly through space;

(iv)   the observed density of stars seen outside the Milky Way(by an observer in or near its plane) depends only on the angle between the line of sight and the plane of the Milky Way; and

(v)   the whole system of stars is arranged in layers parallel to the Milky Way, the density being uniform in each layer but varying from one layer to the next.

This is a reasonably fair summary of Wilhelm's assumptions, although it is questionable how far (i) and (iii) or (iv) and (v) are independent of each other. Encke stated his objections to all five; each objection amounted to an assertion that the assumption might be acceptable as a very broad generalization, but was not useful for the derivation of quantitative results. Encke was denying, in effect, the validity of statistical inference; perhaps he was influenced by the training in strict mathematical rigour given him by Gauss. The final paragraph of the review, though couched in deferential language, was quite damning:

> Struve's great services to the whole of astronomy --admired by no-one more than I-- and which he has crowned with the establishment of an observatory the size of whose instruments as praiseworthy as it is, almost fades against the purposefulness of the whole layout --and in the light of his discerning eye with which the greatest excellence is striven for-- all those things have made it difficult for me to view entirely differently a work to which he has unmistakably dedicated a great love and expense of time and energy. I will readily admit any error, if it is demonstrated to me, but as I see the matter now, it appears to me of importance for astronomy that the assumptions and parallaxes of the Etudes do not get into our astronomical and popular writings.[15]

Three of Wilhelm's letters to Airy mention this book and preserve its author's reaction to these criticisms. Airy mentioned the book in his letter of September 28, 1847, describing his journey home from Pulkovo --perhaps the *Etudes* was one of many presents he took back with him; at any rate he asked for clarification of one or two points, which Wilhelm gave in his reply of October 22/10. In February of the following year, Airy wrote again. He had now read the book thoroughly and devoted two paragraphs to it, mentioning his report to the Royal Astronomical Society . In the first paragraph he wrote:

> I have read Mr. Encke's objections and I think them extremely frivolous. Your able investigation goes on the supposition that though there are considerable irregularities of distribution yet the general distribution may be well expressed by a law. Encke thinks of nothing whatever but the irregularities. I partly think that the parallaxes cannot be so certain as you suppose them... I hope that you will not reply to Encke. [Airy's emphasis][16]

In his reply in April 1848, Wilhelm wrote:

> Now I come to my Etudes. Hitherto I have made no answer to Encke. I am quite of Your opinion that his objections are frivolous and not scientific. It seems to me that he did or would not understand my book. I therefore feel me very much inclined to follow Your advice not to answer at all. That You have given an account of the Etudes to the R.A.S., is an honour done to me by Your kindness, and I express You my sincere thanks for it. I hope to see by next occasion the substance of Your account in the annual Report.[17]

This letter also contains some discussion of Airy's specific questions and of his comment about the unfathomability of the heavens. Wilhelm mentioned the latter finally in December 1848, after he had obviously read Airy's review:

> I have to thank You particularly for the kind manner in which You have mentioned my Etudes in the annual report to the Astron. Society, and I regard it as a spesimen [sic] which has entirely delivered me from answering Encke's frivolous objections. I intend to return once more to the general laws of the distribution of stars in the heavens, in reference to those important materials which Sir J. Herschel's Cape-observations furnish. But I shall not do it before having finished some series of researches on the same subject by application of our great telescope; and therefore the Catalogues of the zone-stars by Bessel and Argelander are completed.[18]

Encke was not Wilhelm's only critic, however, and the latter was perhaps somewhat surprised that John Herschel also criticized him publicly, although in much more temperate terms. Two years after the *Etudes* appeared, Herschel published the first edition of his *Outline of Astronomy*. Even in the tenth edition, published nearly twenty years later and after Wilhelm's death, the younger Herschel still devoted several pages to a discussion of the basic ideas of the *Etudes*. He conceded that Wilhelm's formulae represented the observations well and that his reasoning was without flaw provided "1st, that the level planes are continuous, and of equal density throughout; and 2ndly, *that an absolute and definite limit is set to telescopic vision, beyond which, if stars exist, they elude our sight, and are to us as if they existed not*".[19] (Herschel's italics.) John Herschel believed that his own observations in the southern hemisphere (which, he freely admitted, were not available to Wilhelm when the *Etudes* was written) disproved both assumptions. He saw black areas in the Milky Way on which no stars were projected and supposed that he was looking right out of the system. He also saw areas in which there were bright nearby stars and faint distant ones, but none of intermediate brightness. This, he thought, showed that layers of stars were not everywhere contiguous. Ironically, we now regard both these observations as evidence for Wilhelm's most important suggestion --the existence of interstellar absorption. Wilhelm had proposed a general absorption rather than discrete clouds, but, at that stage in his life, John Herschel had little time for any absorption:

> It would lead us too far aside of the objects of a treatise of this nature to enter upon any discussion of the grounds (partly metaphysical) on which these views [Wilhelm's absorption] rely. It must suffice here to observe that the objection alluded to, if applicable to any, is equally so to every part of the galaxy. We are not at liberty to argue that at one part of its circumference our view is limited by this sort of cosmical veil which extinguishes the smaller magnitudes, cuts off the nebulous light of distant masses, and closes our view in impenetrable darkness; while at another we are compelled by the clearest evidence telescopes can afford to believe that star-strewn vistas lie open, exhausting their powers and stretching out beyond their utmost reach, as is proved by that very phenomenon which the existence of such a veil would render impossible, viz. infinite increase of number and diminution of magnitude, terminating in complete irresolvable nebulosity.[20]

Today, we believe both Wilhelm's and John Herschel's arguments to be wrong, but the latter's triumphed in their day --perhaps because Herschel's reputation stood high not only with astronomers, but with all scientists. From a modern point of view, Wilhelm undoubtedly tried to draw statistical inferences from too few data --to this extent, Encke's criticisms of the values derived by Wilhelm for mean parallaxes were justified. Wilhelm's statistical procedures were lacking by our standards --but great advances have been made in statistical theory in the last 150 years. The memoir in which Poisson introduced the distribution that now bears his name was published only a few years before the *Etudes*. Pogson's modern definition of the magnitude scale -- which would have been a powerful tool for Wilhelm-- came nearly a decade later. Encke, on the other hand, displayed far less understanding of statistical methods and some of his objections were "frivolous". Airy's other stricture, that Encke saw nothing whatever but the irregularities, was also justified. Even though Wilhelm's reasons were wrong, he was right about absorption, and stellar astronomy might have advanced more quickly if his contemporaries had paid more attention to that idea. He made a reasonably good estimate of the size of the part of the Galaxy that he could see, and he was right to emphasize that its true extent is much greater. His idea that the Sun did not lie in the plane defined by the Milky Way (also criticized by John Herschel) is still an open matter. Wilhelm's relatively crude star counts, however, could not have given a definitive answer to that question. Nevertheless, the *Etudes* was the first serious attempt to go beyond William Herschel's studies of the Galaxy. That it was written at all is more important than whether it was right or wrong. It was a necessary preliminary to the work --after Wilhelm's death-- of Charlier, Seeliger, Kapteyn, Karl Schwarzschild, Eddington, and even of Shapley, who finally showed that Wilhelm and the Herschels alike had been wrong in assuming the Sun to be near the centre of the Galaxy.

In all his other work, Wilhelm was a careful observer and painstaking calculator. Even the "calm speculation" of the *Etudes* seems out of character -- at least so Encke appeared to think. Yet this little book develops ideas that Wilhelm was obviously entertaining even when he wrote the introduction to the *Catalogus Novus*. As von Oettingen pointed out in his memorial address on the hundredth anniversary of Wilhelm's birth, every idea in the *Etudes* was present, in germ, in the earlier work.[21] What is more remarkable is that there was so little sign of such a speculative streak in Otto's work or in that of Wilhelm's two astronomer grandsons, Hermann and Ludwig. Only in the latter's son, the younger Otto, did it appear again. Much of *his* work was speculative and severely criticized by contemporaries. He, too, sometimes was right for the wrong reasons: he, too, was a great astronomer.

The *Etudes* was the climax of all Wilhelm's work. Our knowledge of the universe progresses outwards from the Earth. The size and shape of the Earth must be known if we are to determine the size of the solar system, and that, in turn, must be known for us to be able to deduce the distances and luminosities of the stars. Wilhelm knew this well and made fundamental contributions to the first and last links of the chain. His work on the constant of aberration was even a contribution to the middle link, since from that constant the size of the Earth's orbit can be deduced, if the speed of light is known. Flawed as the *Etudes* may be, it was an attempt by a leading astronomer of his day to synthesize his life's work, and to take stock of what he could claim to know of the universe in which he lived. For that reason alone, it deserves our respect.

NOTES

1. F.W. Herschel, 1784-5, Philosophical Transactions of the Royal Society, Vol. 74, pp. 437-451 and Vol. 75, pp. 213-266. (Herschel's papers have been reprinted: see Chap. 3, ref. 7.)

2. F.W. Herschel, 1818, ibid, Vol. 108, pp. 429-70.

3. F.G.W. Struve, 1847, Etudes, pp. 47-8.

4. F.G.W. Struve, 1847, ibid., p. 49.

5. M. Weisse, 1846, Positiones mediae stellarum fixarum in zonis Regiomontanis a Besselio inter -15° et +15° declinationis observatarum, ad annum 1825 reductae et in catalogum ordinatae, auctore M. Weisse. Jussu Academie Imperialis Pertopolitanae edi curavit et praefatus est F.G.W. Struve. Petropoli, 4°.

6. F.G.W. Struve, 1847, Etudes, pp. 67.

7. H.W.M. Olbers, 1826, Ueber die Durchsichtigkeit des Weltraums in Bode's Jahrbuch fur 1826, pp. 110-121.

8. R.J. Trumpler, 1930, Lick Observatory Bulletin, No. 420

9. F.G.W. Struve, 1848, letter of December 31/19 to J. Lee (in the archives of the Royal Astronomical Society and quoted with permission).

10. G.B. Airy, 1847-8, Monthly Notices of the Royal Astronomical Society, Vol. 8, pp. 91-5 (p. 91) (quotations by permission).

11. G.B. Airy, 1847-8, ibid., p. 91.

12. G.B. Airy, 1847-8, ibid., p. 95.

13. J.F. Encke, 1847, Astronomische Nachrichten, Vol. 26, pp. 337-350 (p. 337) (Quotations by permission).

14. J.F. Encke, 1847, ibid., p. 338.

15. J.F. Encke, 1847, ibid., pp. 349-50. I am particularly grateful to Prof. Hadley for help in translating the rather convoluted original of this paragraph.

16. G.B. Airy, 1848, letter of February 18 to F.G.W. Struve (RGO 939 ff. 563-4).

17. F.G.W. Struve, 1848, letter of April 20/8 to G.B. Airy (RGO 939 ff. 567-8).

18. F.G.W. Struve, 1848, letter of December 31/19 to G.B. Airy (RGO 940 ff. 168-9).

19. J.F.W.Herschel, 1868, Outlines of Astronomy, 10th edition (1st edition 1849), Collier and Son, New York, Part Two, p. 710.

20. J.F.W.Herschel, 1868, ibid., p. 714.

21. A. von Oettingen, 1894, Gedächtnissrede, p. 77.

CHAPTER 11

WILHELM'S ILLNESS AND LAST YEARS

In youth and early manhood, Wilhelm had been physically very fit. His father had encouraged him in athletic pursuits when he was a child, and geodetic field-work had preserved the health of the mature man. After the childhood diseases, Otto wrote, Wilhelm had no illness except for occasional trouble with a tapeworm.[1] After Wilhelm returned from his visit abroad in 1853, however, the preparation of *Arc du Méridien* for publication took all his energy, and his way of life became quite sedentary. By this stage of his life, Wilhelm had become a chain-smoker of cigars --his average was fifteen a day and he came to need a cigar in his mouth almost all the time.[2] At first, journeys into St Petersburg for administrative duties and meetings of the Academy at least provided Wilhelm with much-needed fresh air and mental rest, but in January 1855 Paul von Fuss died and Wilhelm even cut down the number of these journeys and

> ...attended sessions of the Academy only quite by exception. What contributed to this circumstance was the fact that, because of conflict of interest, a mood inimical to the Central Observatory had been developing since 1848 among the members of the Academy, which, however, came to light only after the death of Fuss.[3]

Wilhelm never completely resolved this strife with the Academy; it persisted all through Otto's directorship and played an important role at the time of *his* retirement (Chapter 16). With the aid of the Grand Duke Konstantin, however, Wilhelm pressed for new statutes granting the Observatory greater autonomy. This anxiety was a further strain on his health and strength, and by the spring of 1857, when he had sent the first volume of *Arc du Méridien* to press and begun work on the second, Wilhelm "more and more began to complain of tension and fatigue and yearned for a long [period of] relaxation".[4] That summer, he was granted indefinite leave of absence and travelled once again to the west with his wife, three daughters (Alexandra, Emilie and Anna) and his son Karl --who joined them in Berlin. Even now, his work was not left entirely behind. For some time Wilhelm had hoped to measure a great arc of a parallel of latitude --from Valentia on the west coast of Ireland, across the North European plain to Orsk in the Urals; now this project seemed to him to be practicable. "The pursuit of official business was no burden to him", wrote Otto "but only increased [the journey's] joy, since he generally found the friendliest response to the plan."[5] Alexander von Humboldt in Berlin, the French war minister, Vaillant, and Airy in Britain all helped to secure the approval of their respective governments. Work done, Wilhelm was free to travel in the most beautiful parts of Germany, Switzerland and northern Italy. Once again, he saw Argelander in Bonn, and was so impressed by his friend's new assistant F.A.T.Winnecke, that he invited the latter to Pulkovo.

156

Wilhelm was, indeed, mentally refreshed by the journey, but he had overtaxed his strength by more than he knew and was delayed in both Munich and Zurich by outbreaks of virulent boils behind the ears, which, however, he did not regard very seriously. In November, negotiations about the proposed new status for the Observatory reached a crucial stage, and Wilhelm was urgently summoned home. Persuaded by Wilhelm and the Grand Duke, Tsar Alexander II not only agreed in principle to the new statutes, but also considerably increased the Observatory's budget. Wilhelm then plunged back into work on the second volume of *Arc du Méridien*, leaving the house only once in the next four weeks --to thank the Tsar personally. Soon after Christmas, Wilhelm developed catarrh, but welcomed the plausible excuse it provided for him to stay inside. Otto wrote:

> On 26 (14) January 1858, still in a very lively humour, he [Wilhelm] took part in the celebration of my wife's birthday, and nobody suspected that illness was approaching him until, in the late evening, when he wanted to take himself to bed, the company noticed a great swelling had formed in his neck. Early next morning a physician was consulted, who immediately diagnosed the swelling as a virulent carbuncle. Careful treatment and many operations... could not delay the progress of the illness.[6]

So much fluid accumulated in Wilhelm's body that the physicians expected him to die. On 15/3 February, the family gathered round Wilhelm's bed, but a favourable crisis in the illness removed immediate danger. By March 1st, when Otto sent the first news of his father's condition to Airy, he could sound hopeful.

> In the first time the malady offered nothing particularly allarming [sic], but about the middle of the last month the state of the sufferer changed so much that the physicians felt obliged to declare that they despaired of his recovery. Thank God a favourable crisis happened a few hours after this declaration and since that time he is gradually but very slowly advancing to the recovery. We cannot say that at present that he was allready [sic] out of danger, for his feebleness is extreme and the least inadvertency might have serious consequences; but at least it might be said that all symptoms are for a good end of the illness.
>
> A few days ago he asked me to write to you and send his most affectionate remembrances to you and yours...[7]

By early May, however, Otto wrote in a different tone to Airy:

> Your letter of April 27th starts from the supposition that my father has regained his good health and strength. Alas I am sorry to say this supposition has been wrong; on the contrary his state is now subject to very serious apprehensions. The reason for which I have not written you before on this subject is that since 8 weeks we are in a constant agitation with regard to him; several times we have thought him quite out of danger and a few hours later new symptoms appeared that made us fear the worst. It appears that even for his unusual strength the attack of the malady has been to[o] strong. His forces are nearly exhausted, he cannot move without help, he has no appetite, he has lost the memory for all that has passed during the last 20 years; but at present he does not suffer from any pains, neither mental or physical, he is constantly of good spirits and ever of a good humour. The physicians have not yet quite lost the hope for his recovery, for still all

important organs of his body work as they should; but they say that in the most favourable case it might be expected that at least 3 months will pass before he can regain his former strength, and even then it is not certain if he will ever recover his former intellect.[8]

In the same letter Otto describes some of the effects his father's illness had had on him, and adumbrates some of the problems that exercised him throughout the rest of his own life:

From what I have stated you will see that there is more room for apprehensions than for good hope, and will conclude in what condition the family of your friend has been all this time and will be for the next time to come. Most of all perhaps I have to suffer mentally from these circumstances, not only as the eldest son and therewith the present head of the numerous family, but yet more as his locum tenens as Director of the Observatory. You will hardly imagine with what difficulties I have to struggle in the performance of that duty. It is well known to you that since several years nearly all the business matters of the Observatory have been transacted by me, and therefore it might appear strange to you that now I should have to meet particular difficulties on this point. But the difference is that previously I have commonly acted in my father's name and with his authority in the background, [Otto's emphasis] while now I must do it on my own responsibility and, instead of being supported by his authority, I have to fight with all sorts of animosities that his exceptional position has aroused amongst many enviers against himself, his son and particularly against his creation Pulkowa. These animosities appear particularly on the occasion that, in consequence of the greater means given by the Emperor to the Observatory and the extension of our geographical and geodetic operations, new regulations are urgently wanted for Pulkowa. These regulations have been devised last year together by my father and myself; and now every point of them is contested by men who have not the least idea neither of astronomical pursuits nor of the activity of Pulkowa. Until now, I am proud to say, I have successfully resisted to all attempts of ill-will and the heart does not fail me --but I confess I long for a moral support from our friends and from all who esteem my father and appreciate what Pulkova has done for science.[9]

No doubt Otto genuinely longed for moral support from Airy: he had become *de facto* Director of Pulkovo before he was forty years old, and was to remain in charge for the rest of his working life. Four days later, he completed the above letter by adding the latest news of his father's health. Wilhelm was showing signs of both mental and physical improvement, and his son sounded more hopeful: still Otto added "Heaven beware him from a new relapse and grant us a favourable issue of the malady! that is our daily prayer."[10] All through March, Wilhelm had been physically weak and mentally confused. He lived in his first marriage --mistaking Johanna for Emilie-- or even in his childhood. He spoke aloud in Estonian and Lettish --languages he had picked up while surveying in Livland-- Greek, Latin and even Hebrew, which last no-one knew he had learned. He did not remember the events leading up to his illness and sometimes did not even recognize the family. "Also in this respect" wrote Otto

came a sudden crisis. When I came to his bed one morning, he spoke with me as if I were a complete stranger. On my remarking 'Father don't you know me? I am Otto" he first looked at me fixedly, then drawing me to himself with the words "Otto, my old comrade, I

did not know you, that is terrible" he broke out into a stream of tears --an appearance
that we had never before known in him. From this moment the memory for recent times
began to come back, but remained very weak.[11]

In July Otto wrote to Airy again, saying that Wilhelm had much improved now that summer had come, although his memory was still poor. The physicians now hoped for a complete recovery, but recommended that Wilhelm spend the winter in a warm climate. Thus Otto faced at least another six months as Acting Director, of which he said "I cannot say I am much pleased"[12] but at least the summer holidays in which "public affairs repose nearly all" gave him some respite. The letter ends with news of Otto's children, especially of Airy's godson "now nearly as tall as myself"[13] who was making good progress in school --especially in mathematics.

One attempt to reconstruct a modern description of the course of Wilhelm's illness suggests that the so-called carbuncle was probably caused by a bacterial infection. In turn, it released the bacteria, or the toxins created by them, into the blood-stream, causing septicaemia --in popular terms "blood poisoning"-- which spread the toxins to the heart. The gathering of fluid in Wilhelm's body indicates that the heart was not working properly, and suggests myocardial damage, which was shown to be reversible by the sudden recovery called a "crisis" by Otto. The loss of recent memory, which became apparent some time after the reversible heart disease, suggests the "Korsakhow psychosis", a kind of amnesia associated with lesions in certain central regions of the brain. Such lesions might have been caused either by bacterial emboli from the original infection, or more probably by bacterial thrombi from the heart. Wilhelm's continued poor recent memory and inability to work intellectually during the rest of his life could have been residual symptoms after the healing of multiple encephalitis.[14]

Wilhelm was granted leave of absence for an entire year and Otto was confirmed as Acting Director. Once again, Wilhelm set out with his wife and two daughters --Anna and Alexandra-- first to the shores of Lake Geneva then-- for the coldest part of the winter-- to Algeria, back to Switzerland, to Baden-Baden and finally, to Wiesbaden. While they were travelling, the youngest daughter, Anna, was married to Ludwig Cremers, director of a St Petersburg insurance company. Early in September 1859, the rest of the family returned home to the welcome that has already been described (Chapter 7). Although, as that account makes clear, Wilhelm was much restored by his trip, his powers were no longer equal to resuming the direction of Pulkovo. He never completed the second volume of *Arc du Méridien* --Otto and Döllen finished it for him. Neither did he attempt to take up the reins again, although he remained Director in name and a member of the Academy until 1862, when he realized that he was creating difficulties for Otto by continuing in these positions, which he then resigned.

Writing to his daughter Olga Lindhagen and her husband in January 1860, Wilhelm said:

Photographs of Wilhelm and Johanna, c. 1860, copied from a print that used to hang in the home of Olga Lindhagen.

About myself I can upon the whole give you rather good news, especially when I compare my present state of health with that of a year ago. How difficult it was for me to write then, and how easily and steadily I now wield my pen!

Next summer I will not set out on a journey to a foreign watering-place, since --God be praised-- I do not need it.[15]

Instead, Wilhelm invited the Lindhagens to come to Pulkovo. His writing in the letter is indeed firm and clear, and Wilhelm felt well enough to take up some scientific occupation. He wanted to summarize all observations (not only his own) of double stars and worked hard at the task for as long as he was able. Otto wrote of hundreds of sheets in his father's handwriting, deposited in the Pulkovo archives but of no scientific value.[16] Wilhelm's memory for recent events was now so poor that he frequently repeated himself, even on the same page, while in other places he omitted essential information. It became clear, moreover, that he no longer possessed the necessary critical judgment. The kind of work Wilhelm had in mind had to wait some decades after his death until S.W. Burnham produced his *General Catalogue of Double Stars*.[17]

In 1862, Wilhelm and Johanna bought a house in St Petersburg where, thanks to the generosity of the Tsar, he lived without any financial worries,

Photograph of Wilhelm Struve in 1857. Titriumov based a portrait, painted after Wilhelm's death, on this photograph.

surrounded by family and close friends, but no longer a world figure. Apart from persistent catarrh and weakness of the feet that permitted him to take only short walks (a complaint to be echoed by Otto in *his* old age), Wilhelm's health remained good. In summer, he went back to Pulkovo to enjoy the company of children and grandchildren.

Wilhelm's grave at Pulkovo, showing the birches that he had planted in a tree nursery.

In October 1863, six months after his seventieth birthday, Wilhelm celebrated an anniversary that is still vouchsafed to few of us: the fiftieth anniversary of his doctorate. The following August, he was able to take part in Pulkovo's silver jubilee. According to his youngest son, Nikolai, Wilhelm was brought into the great rotunda at Pulkovo, on Otto's arm, to greet his old friends and to inaugurate the festive assembly.[18] He did not long survive this celebration. Back in St Petersburg, early in November, he fell ill with pneumonia. His daughter Emilie, youngest child by the first marriage, and Johanna looked after him in this illness. Shortly after Wilhelm's death Emilie wrote to her sister Olga Lindhagen.[19] Wilhelm had a quiet night on November 7th and appeared to be improving on the 8th (these are Emilie's dates and are Old Style). That day, however, Wilhelm very readily assented when Johanna asked him if he would like to receive the Sacrament. Even on the 9th, Wilhelm appeared better and was visited by several of his family. On the 10th, however, he was crying out in intense pain. Emilie records that the last words he uttered --during this period-- were "Rest! Rest!" (*Ruh! Ruh!*) --the one thing he had scarcely permitted himself throughout his life. At 7 a.m. on the 11th (23rd) November, the family collected around his bedside as Wilhelm breathed his last. His body was buried in the protestant cemetery at Pulkovo: the grave, Nikolai told us, was marked by four birches that Wilhelm himself had once planted in a tree nursery. From the grave, one could look through the trees to the doors of the Observatory.[20]

News of Wilhelm's death soon reached his colleagues in other countries. A telegram was sent to Airy who sent his condolences to the family in a letter dated "1864 November 25 evening".

My Dear Sir,

I received a telegram communicating the death of your respected father, on the 23d, but I could not ascertain with certainty whence it came, for no place was mentioned. Yesterday I notified it to Sir J. Herschel and Mr de la Rue.

I am very much moved indeed by this information, I had known your father from the year 1830 and there was no person of all my acquaintances, in whose knowledge of the subjects, which he took up, and judgment, and kindness of character, I had more complete confidence. I think there is only one person whom I could compare with him on these grounds.

To me he was always very kind and I believe that we never had a single discord of opinions.

I feel the loss almost like the loss of a relative of the family.

Very lately I attended the funeral of Mr. Robert Ransome whom you knew at Ispwich. He was my oldest living friend.

Mrs Airy is gone with my daughter Annot to Wilfrid in the Isle of Wight. She would, I am sure, join with me in all that I have expressed.

I hope that you are advancing successfully in your profession.

I am, dear Sir,

Yours very truly

G.B.Airy.[21]

It is not quite clear for whom Airy intended this letter. He correctly believed Otto might be away from Pulkovo and sent it to "Mr. A. Struve". This could have been Otto's second son Alfred, to whom it would have been reasonable for Airy to address the last paragraph. It might have been meant for Wilhelm's nephew Adolf, but neither for him nor Alfred do the references to "your father" make sense. Moreover, only Otto or Wilhelm himself were likely to have met Mr. Robert Ransome --an engineer who had built instruments at Greenwich for Airy. The normally precise Airy seems indeed to have been so moved by Wilhelm's death that he himself was confused about who would read the letter.

Existing portraits and photographs of Wilhelm show a man of considerable presence and character. Most of them, however, either show him sitting or show only the upper part of his body and do not reveal his imposing height --a family characteristic. This height, his blond hair and blue eyes gave him the appearance (as Otto remarked) of being of typical North German stock.

His height also disguised his unusually large skull "of which, to my knowledge no exact measure remains".[22] His temperament was distinctly choleric, but Otto could recall no occasion on which his father acted rashly while in a bad temper. Wilhelm was never malicious, except in fun, and good-natured to the point of weakness --being always ready to think the best of others. He was generous with money: despite the generosity of the Tsar to him, he never amassed much property. He was also generous in the intellectual help he gave to fellow scientists --for which he claimed no acknowledgement. Before his illness he enjoyed his food, but was no gourmet --except, wrote Otto who enjoyed a good one himself, in the matter of cigars.[23] He had very little artistic sense but a wide knowledge of languages, including classical languages. Otto thought highly of his father's Latin style and regretted the passing of the custom of using Latin for academic writing, since it was so much easier to learn one language than several. Although Wilhelm mastered several modern European languages, he was never fully at home in Russian; his native German remained Pulkovo's working language. He continued, in adult life, to read the Greek and Roman philosophers, but he retained his early dislike of the "hair-splitting" of philosophers contemporary with himself, although some influence from Hegel (whose philosophy was new in Wilhelm's student days) is obvious from this quotation from the *Erinnerung*:

> He [Wilhelm] did not like to discuss religious matters. I do not recall ever to have had a more thorough discussion with him on religious questions, except perhaps from an historical point of view about the origin and development of various confessional differences. He himself describes his religious standpoint in a letter that he wrote in 1810 in answer to an old friend who had urged him to take over [responsibility for] religious instruction. "I cannot teach positive religion because I do not believe it, much as I esteem the Christian religion as the ideal of a 'Volksreligion' and regard its teachings as the teachings of the greatest human being and benefactor. Let one try to teach them natural religion at every opportunity and rouse [in them] the feeling for the Good". He remained true to the philosophy expressed in these words throughout his whole life. Without being permeated by the truth of Christian dogmatics, and on the other hand without imposing his views on others, he remained a faithful adherent of the Christian, and in particular the protestant, Church, practised its teachings in his life, and saw to it that a Christian sense ruled in his family.[24]

Indeed, Wilhelm counted Lutheran pastors among his friends, especially the pastor of the congregation to which he belonged and the Lutheran bishop of St Petersburg who visited him more than once during his illness. The letter quoted by Otto may make Wilhelm's willing acceptance of the Last Sacrament seem somewhat surprising, but the letter was written fifty-four years before Wilhelm died, and Johanna's influence may have changed her husband more than Otto realized. Cleveland Abbe wrote that the text for the sermon at Wilhelm's funeral was "God is love and who abides in love abides in God and God in him"[25], but perhaps Wilhelm would like to be remembered by his favourite motto, a line of verse from Horace: *Aequam memento rebus in arduis servare mentem*, which can be translated "Remember, when life's path is steep to keep your mind even".[26]

Wilhelm's reputation as a great astronomer (his descendants still call him *the* great astronomer) is secure. He was not a great innovative genius --he

Monument to F.G.W. Struve in Tartu (not far from the old Dorpat Observatory) designed by Udo Ivask.

clearly and gladly acknowledged the superiority of William Herschel in his major fields of interest (except geodesy). If the word "genius" is applied to Wilhelm at all, perhaps it should be in the sense defined by his contemporary Thomas Carlyle: the "transcendent capacity of taking trouble, first of all". The manifold activity of Wilhelm's early life speaks for itself, and the testimony of men like Olbers and Airy to his skill as an observer cannot be ignored. In his time, Wilhelm was regarded as one of the leading astronomers of the day and it was not just filial piety that made Otto write "The history of science will still speak of him after centuries".[27] Over a century after his death and nearly two after his birth, we do indeed speak of him and --more to the point-- we still use his observations. *That* astronomers will do as long as they are interested in double stars, for Wilhelm's observations are the best of his time and their value increases with age.

A year before Wilhelm died, Otto played a considerable part in the foundation of the *Astronomische Gesellschaft*. In that society's new *Vierteljahrsschrift*, Argelander wrote an obituary of his good friend Wilhelm Struve. Its closing words provide Wilhelm's best epitaph:

> As a man, Struve was one of the noblest, full of love for his fellow-men, always ready to help, where he could help, mild in his judgment of others, strict with himself in the fulfillment of his duties, charming in company, a true husband, a loving father and a sincere friend. As a scholar he was distinguished by acumen, consistency and perseverance of the highest degree, a rare observing talent and an extraordinary mobility of mind, that enabled him often to follow a number of considerably heterogeneous tasks simultaneously. He left posterity a rare example of human perfection. May his ashes rest in peace![28]

NOTES

1.   O.W. Struve, 1895, Erinnerung, pp. 63-4.
2.   O.W. Struve, 1895, ibid., p. 75.
3.   O.W. Struve, 1895, ibid., p. 60.
4.   O.W. Struve, 1895, ibid., p. 67. (I have read "Anspannung" for "Abspannung").
5.   O.W. Struve, 1895, ibid., p. 67.
6.   O.W. Struve, 1895, ibid., p. 69.
7.   O.W. Struve, 1858, letter of March 1 to G.B. Airy. (RGO 948 f. 503).
8.   O.W. Struve, 1858, letter of May 6 to G.B. Airy. (RGO 948 ff. 504-5).
9.   O.W. Struve, 1858, ibid.
10. O.W. Struve, 1858, ibid.
11. O.W. Struve, 1895, Erinnerung, p. 70.
12. O.W. Struve, 1858, letter of 23/11 July to G.B. Airy. (RGO 948 f. 507).
13. O.W. Struve, 1858, ibid.
14. This paragraph is based on a letter of August 25, 1985 (in English) from Prof. P. Sourander (Emeritus Professor of Physiology in the University of Göteborg) to Dr. Nils Lindhagen.
15. F.G.W. Struve, 1860, letter of January 16/4 to D.G. Lindhagen and Olga Lindhagen. The letter is the property of Nils Lindhagen who has provided a copy and a translation (used here in its essentials).
16. O.W. Struve, 1895, Erinnerung, p. 72.
17. S.W. Burnham, 1906, A General Catalogue of Double Stars within 121° of the North Pole (in two parts), Carnegie Institution of Washington.
18. Nikolai Struve, 1915, Deutsche Monatschrift für Russland, (see Chap. 9, ref. 3).
19. Emilie Struve, 1864, letter to Olga Lindhagen written a few days after Wilhelm's death. The letter is in the possession of Nils Lindhagen who has furnished me with excerpts.
20. Nikolai Struve, 1915, see ref. 18 and Chap. 9, ref. 3.
21. G.B. Airy, 1864, letter of November 25 to A. Struve. (RGO 951 S38).
22. O.W. Struve, 1895, Erinnerung, p. 74.
23. O.W. Struve, 1895, ibid., p. 75.
24. O.W. Struve, 1895, ibid., pp. 76-7.

25   Cleveland Abbe, unpublished autobiographical fragment, p 57 (Library of Congress, Cleveland Abbe Papers, Container 32)

26   This translation of the line from Horace (Odes II 3 1) may be found in the <u>Oxford Dictionary of Quotations</u> (2nd edition 1953, reprinted 1968) Oxford University Press, p 259, No 3

27   O W Struve, 1895, <u>Erinnerung</u>, p 35

28   F W A Argelander, 1866, <u>Vierteljahrsschrift der Astronomischen Gesellschaft, I Jahrgang</u>, pp 31-52 (p 52) (Quoted with permission from the <u>Astronomische Gesellschaft</u>)

CHAPTER 12

THE TRANSITION

Although Otto had been officially Director of Pulkovo for two years
before his father died, (and *de facto* for much longer) the removal of the
founder's physical presence must have seemed the end of an era to all those
connected with the Observatory. From our vantage point, Wilhelm's death
divides the Observatory's first fifty years even more clearly into two parts,
since it occurred almost exactly halfway through the time of the Struve
family's hegemony. The late 1860s and early 1870s thus became a time of
transition in the life of the Observatory, and this transition was reflected in
the circumstances of Otto's personal life.

The new statutes, over which Wilhelm and Otto had worked so hard
together, came into force with Otto's directorship and permitted some increase
in staff. In addition to the four senior astronomers, there was provision for
two assistant astronomers and two computing assistants, while the Observatory
could also supply space and free lodging for unpaid supernumerary
astronomers. In addition, the Observatory was authorized to employ a medical
doctor whose first duty was to attend to the other members of the staff and
their families. Previously, the staff had had to seek the services of a doctor 14
versts away, or of the head physician of Tsarskoe Selo hospital.[1]

The last appointment that Wilhelm had made to the staff had been that
of Winnecke, who arrived after his patron had been taken ill. Otto appeared to
share his father's good opinion of this young man, for in 1862 he wrote of him
to Thomas Maclear (of the observatory at the Cape of Good Hope) as "since 4
years one of my assistants and a most talented astronomer".[2] From a letter
written to Airy two years later, we learn of Winnecke's engagement "to a young
lady of Dorpat"[3] and the subsequent marriage connected this young astronomer
to the Struve family. Apart from Otto himself, the only member of the original
staff left at Pulkovo by this time was Döllen, by this time a senior astronomer
--as were also Winnecke and A. Wagner (the fourth position was vacant). Peters
and Sabler had long since left, and Georg Fuss, who had left to become the
Director of the Vilna Observatory (modern Vilnius), had died in 1854. Georg's
son Victor, however, had become a supernumerary astronomer at Pulkovo.
Lindhagen had returned to his native Sweden a few years after his marriage to
Otto's sister, Olga, but a compatriot of his, J.A.H. Gyldén --who also later
returned to a distinguished career in his own country-- came to Pulkovo as one
of the new assistant astronomers. The other assistant was Captain Smyslov (the
"Smythlove" of Chapter 7) who, however, was more concerned with the geodetic
side of the Observatory's work.

Emilie (née Dyrssen) and Otto, photographed in 1863.

Around the time of his father's death, Otto too was ill and took sick leave for the winter of 1864-5, in order to recuperate in the warmer climes of southern Europe. He spent much of the time in Italy, and during this and subsequent visits, he cultivated his friendship with the famous Italian astronomer G.V. Schiaparelli, who had visited Pulkovo in 1860. Otto left Winnecke in charge back at home, and the future must have seemed bright to this young man when he found himself Vice-Director of one of the world's great observatories before his thirtieth birthday. His subsequent life was unhappy, however, for he twice suffered mental breakdowns that incapacitated him for several years. The first came in this very winter of 1864-5, and is usually attributed to overwork from his student days onwards. An illness contracted during an eclipse expedition to the Himalayas appears to have played some part in his mental illness, however, and a nervous temperament is suggested by one account which refers to his suffering an attack of acute terror one night while observing in the dome.[4]

Winnecke was in charge when Cleveland Abbe arrived in Pulkovo, and received Abbe in a manner that the latter described as "cool". Abbe was quick to use this term of anybody to whom he did not warm himself, but he soon realized Winnecke was ill, and told his parents in his first letter home:

> From him [Winnecke] I learned first to realise how completely shut out from the world we are, and as he is in very poor health I suppose he made the picture as gloomy as need be.[5]

Despite this illness, Winnecke invited Abbe to dine, but his condition rapidly grew worse and there were fears that he had contracted typhus. By the end of March, Abbe wrote of him again:

Giovanni Virginio Schiaparelli.   (Reproduced by permission of the Royal Astronomical Society)

> Dr W I saw on Sunday for the first time since he took to his room It's very sad to see so
> fine a mind of such unusual promise so soon prostrated He hardly said a word and will
> not think or speak more than he can help I do not think this is <u>entirely</u> the result of too
> much study [6]

Two weeks later, on April 13, Abbe wrote that Winnecke, still depressed, and maintaining (truly enough) that he would never return to Pulkovo, had left for Germany. Rest and therapy there led to an apparently complete cure and Winnecke became Director of the Strasbourg Observatory --after the Franco-Prussian war had made Alsace-Lorraine a part of Germany. He even became Rector of the University, but the death of his eldest son in 1881 triggered a new bout of depression which completely incapacitated him until his own death in 1897.

Otto did not seem to think that Winnecke's illness was a sufficient reason for cutting short his own sick-leave, even though he wrote about it to Airy, who considered the matter important enough for a special letter of sympathy.[7] Abbe was thus left with little to do until Otto's return in September 1865. He quickly made friends with Victor Fuss, and these two were joined -in university vacations-- by Otto's youngest half-brother, Nikolai, later a gymnasium teacher in Riga. Abbe was also made welcome by Gyldén and Döllen, and felt something in common with Wagner, whose wife (daughter of the German astronomer Hansen) was sister-in-law to the American writer Bayard Taylor. Abbe was gratified to find his new friends on *his* side in the American Civil War:

> Dr D[öllen] is very much interested in American affairs and knows much about them He
> says truly that all look with much anxiety to America and hope that the South with
> Secession Slavery and Rebellion will be put down For the conviction of all Europe is that
> slavery stops progress The Emperor Alexander Nicolaivitch is a strong persevering man
> and means to do right in spite of his nobles and court 8

Some of these friends, however, --it is not clear which, except that it is inconceivable that Döllen should have been among them-- sowed doubts in Abbe's mind about the sort of reception he would receive from Otto. Later, when Abbe hoped that he might both stay permanently in Pulkovo and marry into the Struve family, he found relations with some of these colleagues became strained:

> The position that Mr Otto Struve wishes to secure for me is combatted by others on the
> ground that I am a foreigner and by others on the ground that I wish to marry his
> daughter or some other member of the family and that in this way he is seeking to
> establish still firmer the hold of the Struve family upon the Observatory Much is said of
> Pulkovo as a family affair and not so entirely devoted to Astronomical Science as it should
> be The Elder Struve by his universal kindness and self control kept all in harmony --but
> Otto S is not so successful

There was indeed a tangled web of matrimonial alliances at Pulkovo. Two astronomers, Döllen and Lindhagen, became Wilhelm's sons-in-law, and a third, Winnecke, was connected to the family by marriage. Later, another Swedish astronomer, Nyrén, married into the family of Wilhelm's first wife.

Wilhelm Döllen photographed in St. Petersburg by H. Steinberg.

Like Abbe himself, most if not all of these came to Pulkovo as young bachelors. Their social life centred around the Director's house, and many of the young ladies they met inevitably were more or less closely related to the Director. It was natural enough, therefore, for young astronomers to marry into the family but there is no evidence that anyone was appointed just because he had done so, or hoped to do so. It was equally natural for the others to feel that they were not in the "inner circle" and to complain about "Poulkova as a family affair". The continued productivity of the Observatory under Otto's leadership is surely sufficient evidence that it was "devoted to Astronomical Science" but it did seem to many that he regarded Pulkovo as a family fief. At the time of Abbe's visit, there was probably sufficient uncertainty about the future to provide a fertile ground for the kind of grumbling he records. Otto himself wrote to Abbe (in English) shortly after the latter's return home:

> While I write this you will probably have reached your home since more than a week and your stay at Pulkowa will appear you like a dream I am afraid this dream might not be throughout a pleasant one; for certainly these two years that you have spent with us were the least favourable by exterior circumstances that Pulkowa ever has witnessed. My father's death, my own illness those of Winnecke, Döllen etc; together with many other displeasing circumstances, were not qualified to make your stay here a pleasant one. With

all that I hope you will have taken with you a friendly recollection of the inhabitants of the Observatory and be satisfied with the general character of our working and proceeding.[10]

It appears that Otto's wish was fulfilled. Abbe later wrote an article "Dorpat and Pulkovo"[11] on which Otto commented "you have made Pulkovians, and especially the Struve family very happy by this paper".[12]

During this transition period, Otto was still engaged in his own scientific work. Even before his father died, he had been concerned about the possibility of systematic personal errors in his measurements of double stars. To try to estimate these, he made "artificial double stars" by inserting ivory plugs in a black board. He could measure the apparent separations of these with a telescope and micrometer, and also directly. He came to believe that he was systematically overestimating separations, and carefully checked --by measuring the same pair of stars at quite different elevations-- that his conclusion was not a result of the differing orientations of the telescope as used to measure real and artificial doubles. He published a preliminary account of this work as early as 1860[13], but letters to Airy and Abbe in 1866 show that Otto was still working on the problem and preparing a detailed memoir for the St Petersburg Academy. Personal errors in double-star measurements have long remained a vexed question. Most double stars have orbital periods of decades or even centuries, and their orbits must necessarily be determined from measurements made by several different observers. There are consistent differences between observers, some of whom are more self-consistent (and therefore reliable) than others. Otto and his father are among the most consistent and reliable observers ever, having been exceeded in these respects only by people who have developed completely impersonal methods in the last few decades. If all observers had checked themselves as carefully as Otto attempted to do, personal corrections might be applied to all their results before an orbit is computed. In fact, most of those who compute orbits prefer to be selective in the observations that they use, relying on well-known and trustworthy observers, such as the Struves, whenever possible.

Otto also became interested in astronomical spectroscopy at about this time. Later astronomers have accused him of neglecting the development of astrophysics at Pulkovo and he was certainly more at home in traditional positional astronomy. He did take the first steps in beginning astrophysics there, however, bringing another Swedish astronomer, Hasselberg, to Pulkovo, in the late 1860s, for that specific purpose. In May 1868, he wrote both to Airy and to Sir William Huggins (the British pioneer of stellar spectroscopy) about his spectroscopic observations of the aurora:

A fortnight ago I had twice a good occasion to regard the Aurora Borealis with the spectrum apparatus. Its light is monochromatic. The line was situated on the boundary of yellow and green, a little more in the green. On Kirchhoff's scale it would correspond with the number 1259.[14]

This paragraph was a postscript to a longer letter (in English) concerned with relations between the Royal Astronomical Society and the newly formed *Astronomische Gesellschaft*. It shows that Otto's first measurements of the auroral line (published somewhat later[15]) were made at about the same time as the work of Ångström --usually considered the first to measure the line-- which was published in 1869.[16]   Ångström began his observations in late 1867, however, and Otto's nearly simultaneous work may not have been completely independent, since he could conceivably have heard about Ångström's from his brother-in-law Lindhagen, now back in Sweden.

Kirchhoff's scale is quite arbitrary, and the only way to convert Otto's measurement into modern units is to look up Kirchhoff's original papers,[17] from which it can be calculated that 1259 corresponds to approximately 5553 of the units what we now name after Ångström.   The modern value of the wavelength of the auroral line is 5577 A. Otto's value was neither the best nor the worst of the early measurements, and he himself assigned an uncertainty of 10-15 Kirchhoff units to his result.

Otto was developing his contacts with American astronomers about this time, and for a while he used Abbe as a channel of communication. Later he formed a close friendship with Simon Newcomb, who superseded Abbe as Otto's foremost American correspondent. The first letters to Newcomb were formal and in English.   In 1873, however, Otto found that Newcomb could read German and switched to that language --as he did with other correspondents- in order to save time in his letter-writing. The letters also began to be more friendly, as is shown by the evolution of the salutations, from the formal "Hochgeehrter College" to the friendly "Theurster Freund" used in 1879 and subsequently. The first surviving letter to Newcomb, written in April 1867, makes quite clear what scientific topic brought the two men together:

> Dear Sir,
>
> I am most happy to hear that you will undertake the deduction of the most probable value of the solar parallax from the observations made on Mars in 1862. Though the last news about Dr. Winnecke's health are more promising for his ultimate recovery, still there is hardly a chance that he now will be able to apply himself again to that task... At present our forces are hardly adequate to our ordinary daily work; therefore we cannot but feel very much obliged to you, for your intention of executing the mentioned work.[18]

This promising start to the friendship was nearly wrecked, however, when --as Otto thought-- Newcomb neglected the courtesy of sending a copy of his results to Pulkovo. Otto, clearly displeased, wrote to Abbe for help:

> With reference to a paper, may I impose on your friendly mediation; I mean Prof. Newcomb's work on the parallax of Mars. As you know, we are specially interested in this work, so we had fully expected that Pulkowa would not be put in the position of learning the result first through Paris and still less that we would have to wait a year and a day to obtain the paper itself, if indeed it will be sent at all. If, therefore, you could get hold of a copy of this paper for us, that would be very nice for me.[19]

Abbe acted quickly: he found that Newcomb had sent no fewer than five copies of the paper, and he soon arranged for another to be on its way to Pulkovo. Somewhat mollified, Otto wrote to Abbe:

The copy of his [Newcomb's] paper sent earlier has even now not been received here. I understand from your letter, however, that it was addressed to the Academy of Science, so I will enquire there once more. It is not quite impossible that it will be found there, stuck in some corner.[20]

None of the five copies sent originally ever arrived.[21]   The slowness and unreliability of mail between Russia and other countries were constant complaints of both Wilhelm and Otto. Often their letters contain suggestions for avoiding the use of ordinary mail. Colleagues sending books, in particular, were advised to send them to the nearest Russian consul, who could forward them by diplomatic mail. After this temporary misunderstanding, Newcomb and Otto became regular correspondents, exchanging letters at least once --and often several times-- each year from 1873 until Otto's death in 1905.

Another Swedish astronomer, Nyrén who --as already mentioned-- became connected with the Struve family by marriage, joined the staff of Pulkovo in 1868. Although he returned to Sweden on his retirement in 1908, he spent virtually the whole of his active career in Pulkovo. His first task was to take up the work on nutation that Abbe had left unfinished shortly before. Newcomb was interested in this work, and Otto forwarded a copy of Nyrén's treatise in 1873. It contained the first evidence --that Otto was not entirely willing to accept-- for a variation in the altitude of the celestial pole. Nyrén's next task was to redetermine the constant of aberration that had been determined in Pulkovo by Wilhelm himself. Since this provided a means of estimating the distance of the Sun from the Earth (Chapter 10) it was also of interest to Newcomb --especially after the failure of the observations of the transit of Venus in 1874 (see the next Chapter). Nyrén used Wilhelm's prime-vertical instrument. The observations necessarily took some time, but in March 1879, Otto wrote:

Nyrén has recently concluded his series of observations with the transit instrument in the prime vertical and everything seems to indicate that he will obtain a nearly identical result with that derived by my father. Since, however, Nyrén's observations extended only to four stars... which at the same time, indeed, is more numerous than my father observed, so I have charged the newly arrived Adjunct Dr. Backlund to begin without delay a new series of determinations from as many other stars as possible (he has already started the work). By the spring of 1881 can also this series be very completely closed. Naturally, through the same, your determination of the solar parallax by comparison with the intended direct determination of the velocity of light will gain still greater weight.[22]

Backlund was yet another Swedish astronomer who came to work at Pulkovo. Otto's first (and probably juster) opinion of him was a very high one, but it was to change drastically. Some six months later, however, Otto was much less optimistic about the aberration work. His letter of October 1879 is smudged and hard to read, but it appears to say that, although Nyrén had found a mean value for the aberration constant close to Wilhelm's, individual observations showed a periodic term that was presumably of instrumental origin. Indeed, it was found that the flooring of the observing room had moved and was touching one of the pillars supporting the instrument -- rendering the whole two-year series of observations unreliable. Otto concluded "I have ordered the necessary repairs and, in a few days, Nyrén will begin a

Magnus Nyrén, photographed in St. Petersberg (W. Clasen).

new series of observations for the same purpose".[23] Poor Nyrén! By April 1880,
however, Otto reported that Nyrén had been observing twenty stars since
December and that *this time* he would reduce the observations after each night,
so that no unexpected errors could pass undetected again and thus vitiate the
whole series. The work was finally completed in 1882 and Otto then wrote to
Newcomb:

> Nyrén's observations give, as it appears, an aberration constant <u>several</u> hundredths of a
> second <u>greater</u> than that which my father found, however his calculations are still not
> definitively finished [25]

Another concern of Otto's which began in the transition period and, like
the above observations, extended long afterwards was the sale of his personal
library. He had tried to sell it to the Russian government while Abbe was still
in Pulkovo, but the plans were frustrated by a change in the ministry. Otto
and Abbe must have discussed this matter, and all Otto's subsequent efforts
were directed towards selling it in the United States. In the first letter that he
wrote to Abbe after the latter's return, Otto said:

> Do not forget, if an occasion offers itself, that I should be rather glad to sell my library
> About 2000 volumes and 2000 dissertations, price 5000 dollars [26]

Otto's eldest son, Wilhelm, prepared a catalogue of the library, and when Abbe was appointed director of the Cincinnati Observatory, Otto tried to persuade him to buy it for his new institution:

> As I have said to you in passing before, owning this library is frankly a burden to which I do not have the time to pay sufficient attention, besides I always come into conflict between parcels that are sent to me personally and to the Observatory. In fact, I have little use for it since each book is at my disposal in our Observatory library at any time. For that reason it would be very nice for me if I could sell it. Two years ago I indicated to you a price of $6000, and I won't change that now, although in the meantime many valuable additions have come in, particularly through the completion of significant series.

> It would please me very much if you yourself would acquire this collection for the Cincinnati Observatory. It contains nearly everything important which the practical astronomer requires today if he is not to be impeded in his research by a lack of literary reference aids and, with regard to all the more significant publications in the last half century, is practically complete. In such a place as Cincinnati, where, one might suppose, not very much astronomy is to be found in other libraries, such a "select library" [Otto's own English phrase in a German letter] must be of the greatest significance for the Observatory itself and for its work.[27]

Cincinnati, an observatory founded by public subscription, just did not have $6000 to spend, and Abbe -who did not stay there long-- failed to rise to the bait even when Otto offered to continue to send to the purchaser copies of all the periodicals he received for the rest of his life, at no extra cost. When he began to correspond with Newcomb, Otto enlisted his new friend's help and the founding of the Lick Observatory raised some hopes that his library would find a home there. Otto wrote to Holden[28] who referred the matter to the Newberry library in Chicago. The price had now become $9000, and Otto wanted the capital to provide himself with a pension. This institution also declined the offer, and not until May 1892 was Otto finally able to write to Newcomb:

> It will interest you to learn that I have sold my library, in fact to Columbia College [now Columbia University] in New York. Even if I have not obtained quite the price for it that I considered justified, nevertheless I am satisfied with that which was offered, since with my presently unsettled wandering life [see Chapter 16] the library had become flatly a burden. The College, in any case, made a very good purchase and I am happy that my collection will be in good hands there.[29]

Otto's last link with his father was broken in August 1867 by the death of Wilhelm's widow Johanna. Just over a year later, in a curious parallel with Wilhelm's life, Otto's own wife Emilie died. He described the circumstances to Abbe in reply to the latter's letter of condolence:

> And this blow caught me, in fact, quite unprepared. We had this year spent so happy a summer, such as will come no more. My own condition had much improved, as a result of various happy events that cheered the mind and perhaps will turn out to be a genuine means of cure, and my wife also seemed stronger than in earlier years. Then suddenly she fell sick of pneumonia, without that itself being any immediate cause for alarm, and in 5 days she was no more. Oh! it was, it is even now still very hard to bear!

My daughter Mary manages my housekeeping now; nevertheless my sister-in-law, Winnecke's mother-in-law, is with us for this winter, and genuinely helps us to bear the burden. In this way care is taken that my household affairs have not appreciably changed.

But the real soul of the house is missing![30]

Mary did not leave home until her own marriage in 1873, but her brothers Wilhelm and Alfred were already looking for positions. Otto's letters to Abbe show him to have been a proud father:

...Wilhelm has at last the prospect of a steady job. He is designated as Secretary to the St Petersburg Statistical Committee, and hopes to be able to take up this position already in the next few days. Alfred will finish his studies in the Mining Academy next May, and he likewise will also enter the civil service.[31]

Some months later, Otto wrote:

...my second son, Alfred, is now fully launched into practical life. He has passed his examination as a mining engineer very well, and already 8 days ago entered a position in which the establishment of a coal-mining factory in the Ryazan Government has been assigned to him under very satisfactory conditions. My other children are presently all here for the holidays. It is hay-harvest time here, and that is a well-known great pleasure for the youngsters.[32]

Amongst those youngsters were the two future astronomers Hermann and Ludwig --then 14 and 11 years old, respectively. The progress of the older sons continued; in January 1871, Otto wrote of them:

Wilhelm...now in an honourable position as statistician to the Central Committee and Alfred is technical director for the exploitation of rich coal seams that he discovered himself in the Government of Ryazan.[33]

As his children were leaving home, Otto completed the transition in his personal life by marrying again, in August 1871. His new bride was Emma Jankowsky (the now usual spelling, although the name is "Jankoffsky" on the formal announcements of the wedding). The old family friend Airy was among the first to be told the news before the wedding:

My dear old friend,

I hope the news forwarded by the enclosed card will please you...

Let me also recommend my dear Emma to your benevolence and friendship... In about six weeks our wedding will take place and then we shall immediately make a tour through part of Europe. On this occasion I will not come to England, but perhaps in a year or two, whenever I shall go abroad again, I should like very much to bring my wife over with me and show her my beloved English home.

Every truly yours,

Otto Struve.[34]

Two years later, Emma gave birth to Eva, the only child of the marriage. Much younger than her half-brothers and sisters, Eva was destined to spend her early adult years looking after her elderly parents. Later she lived in Berlin until her own death in 1954. Surviving both world wars, she became a link between Otto's generation and members of the family still alive today, who knew her personally. She also played a part in bringing Otto's grandson and namesake to Yerkes Observatory, in the United States, in 1921.

NOTES

In this and all subsequent chapters, the letters to Simon Newcomb that are cited are all in the Simon Newcomb Papers at the Library of Congress, in container number 40

1   O W Struve, 1865, Uebersicht an die Thätigkeit der Nicolai-Hauptsternwarte während der ersten 25 Jahre ihres Bestehens St Petersburg I have used an unpublished translation by K Krisciunas as a source for the information

2   O W Struve, 1862, letter of May 20 to T Maclear (RGO 15/54 f 132)

3   O W Struve, 1864, letter of March 19 to G B Airy (RGO 951 ff 689-90)

4   E Hartwig, 1898, Monthly Notices of the Royal Astronomical Society, Vol 58, pp 155-9

5   Cleveland Abbe, 1865, letter of January 2 and 4 (New Style) to his parents (Library of Congress, Cleveland Abbe Papers, Container 2)

6   Cleveland Abbe, letter of February 1 to his parents (Loc cit Container 2)

7   G B Airy, 1865, letter of March 27 to O W Struve, in reply to Struve's letter of March 11 (RGO 951 ff 708-9)

8   Cleveland Abbe, 1865, see ref 5

9   Cleveland Abbe, letter of May 30/June 11 to his father (Library of Congress, Cleveland Abbe Papers, Container 3)

10   O W Struve, 1866, letter of November 29 to Cleveland Abbe (Library of Congress, Cleveland Abbe Papers, Container 3)

11   Cleveland Abbe, 1867, Dorpat and Pulkowa (see Chap 6, ref 10)

12   O W Struve, 1869, letter of July 8 to Cleveland Abbe (Library of Congress, Cleveland Abbe Papers, Container 4)

13   O W Struve, 1860, Monthly Notices of the Royal Astronomical Society, Vol 20, pp 341-4

14   O W Struve, 1868, postscript to a letter of May 12 to Sir W Huggins (Royal Astronomical Society Archives, quoted by permission)

15   O W Struve, 1869, Beobachtung eines Nordlicht-spectrum, Bull scientifique publ l'Académie Impériale des Sciences St -Pétersbourg, Vol 13, pp 49-50

16   A J Ångström, 1869, Annalen der Physik und Chemie, Vol 137 (=ser, 5, Vol 17), pp 161-3

17   G R Kirchhoff, 1861 and 1862, Physikalische Abhandlungen der Königlichen Akademie der Wissenschaften zu Berlin, pp 63-95 (1861) and 227-240 (1862)

18   O W Struve, 1867, letter of April 15 to Simon Newcomb

19   O W Struve, 1868, letter of December 3 to Cleveland Abbe (Library of Congress, Cleveland Abbe Papers, Container 4)

20   O W Struve, 1869, letter of March 19 to Cleveland Abbe (loc cit , Container 4)

21   O W Struve, 1869, letter of July 26 to Simon Newcomb

22   O W Struve, 1879, letter of March 30 to Simon Newcomb

23   O W Struve, 1897, letter of October 24 to Simon Newcomb

24   O W Struve, 1880, letter of April 24 to Simon Newcomb

25   O W Struve, 1889, letter of August 22 to Simon Newcomb

26   O W Struve, 1866, letter of November 29 to Cleveland Abbe (see ref 10)

27.   O.W. Struve, 1868, letter of December 3 to Cleveland Abbe   (Library of Congress, Cleveland Abbe Papers, Container 4).

28.   O.W. Struve, 1891, Letter of November 19 to E.S. Holden (Mary Lea Shane Archives of Lick Observatory).

29.   O.W. Struve, 1892, letter of May 17 to Simon Newcomb.

30.   O.W. Struve, 1868, see ref. 27.

31.   O.W. Struve, 1868, ibid. ˇ

32.   O.W. Struve, 1869, see ref. 12.

33.   O.W. Struve, 1871, letter of January 22 to Cleveland Abbe (the date could be 1870; Otto's writing is not clear). (Library of Congress, Cleveland Abbe Papers, Container 6).

34.   O.W. Struve, 1871, letter of June 29/17 to G.B. Airy, (RGO 952 f. 614).

CHAPTER 13

THE COMPANION OF PROCYON AND
THE TRANSITS OF VENUS

Without doubt Otto was a careful and cautious observer: the reliance that is still placed on his double-star measures is sufficient testimony of that, but we also have the evidence of contemporaries such as Piazzi Smyth who saw him at work (Chapter 7). If still more evidence is needed, the work on artificial double stars, or Otto's requiring Nyrén to repeat the aberration observations --both described in the previous chapter-- should provide it. Even the best observers make mistakes, however, and Otto was for a while misled into believing that he had made an important discovery in his own field of double-star observation.

Bessel's discovery of irregularities in the proper motions of the two bright stars Sirius and Procyon has already been mentioned in Chapter 10. Although, as late as 1847, Wilhelm expressed doubt about the reality of these irregularities, Bessel confidently ascribed them to the presence of unseen companions of each of these stars and he was dramatically vindicated in 1862 (after his death) when the American telescope maker, Alvan Clark, detected the companion of Sirius while testing a new telescope. The nature of this companion was not fully understood until the twentieth century, but it turned out to be the prototype of a whole new class of stars --the white dwarfs. Confirmation of Bessel's prediction about one star strengthened belief that he had also been right about the other, and many people began to look for a companion to Procyon. We now know that this star, too, is a white dwarf: it is a curious coincidence of constellation mythology that both dog stars should have similar companions. White dwarfs are very small and very faint. It is particularly difficult to see them with certainty when they are close to a bright star. To succeed in detecting the companion of Procyon, the would-be discoverer had to have access to a large telescope and should, by preference, be experienced in double-star observations. Naturally, Otto attempted this discovery and, in 1873, believed he had made it. He published his discovery in that year[1], complete with measurements of the companion's relative position, and he published further measurements the following year.[2] Several of Otto's letters to Newcomb refer to this topic: about this time, the latter had gained access to the new 26-inch refractor at the U.S. Naval Observatory in Washington, and since *he* now had the world's largest telescope, he hoped to discover Procyon's companion himself. Perhaps this fact stimulated Otto to make his attempt in 1873; at any rate, in June of that year a little friendly rivalry crept into his letter to Newcomb:

> You will perhaps already have heard that I believe I have stolen a march on your new
> refractor in the discovery of Procyon's satellite. I have observed on two evenings in

Two photographs of Otto: left in Stockholm (Johannes Jaeger) and right at his desk in St. Petersburg (W. Clasen).

March, in excellent seeing, a faint point of light, 12" distance from Procyon following it in the parallel [of declination]. As soon as the seeing became only a little unsteady, however, the impression vanished, on which account I am uncertain whether I had allowed myself to be deceived by such an image. It would give me great pleasure if your new giant telescope confirms the discovery next winter. --A little article on this object will probably appear in the next number of <u>Monthly Notices</u>.[3]

Newcomb failed to confirm the discovery before the end of the year, however, and in January 1874 Otto wrote again:

As to Procyon, I have still not got round to it this winter, and probably nearly a month must go by before we may count on seeing as good as is required for this object. I am, in fact, anxious to a high degree that the object seen in the previous year should show itself again, and still more that you, with the undoubtedly superior power of your 26-inch, should forestall me in the confirmation. In any case, allow me only to remark that my opinion [is] that the greater optical power and in itself the higher precision of images is not the decisive [factor] in this matter. The principal role in this is played by the steadiness of the seeing in the atmosphere; but it then, perhaps, depends on a fortunate relationship between the quantity of light emitted by the primary star and the sensitivity of the eye of the observer. Because of the great light grasp of your gigantic objective compared with ours, it might be recommended that the attempt to see the small companion be undertaken in still bright twilight, soon after sunset At least that's how it works for me with the companion of Sirius, I can see that clearly only in bright twilight --I will in any case admit to you that, in respect of the reality of the object seen by me last spring, I am more nearly sceptical because it has still to be seen through my other telescope It is certainly not an illusion, however, since as well as I, two colleagues have seen it clearly and simultaneously on at least two evenings It also cannot have been in the telescope, since otherwise a change-over to the lower position must show the object on the opposite side [The handwriting is here hardly legible and the meaning is correspondingly uncertain and obscure] I hold it possible that on several evenings peculiar and nearly similar reflections in the atmosphere Or do we here have to do with a variable star?4

Otto continued to discuss this problem in his letters for another two years. In the spring of 1874 he was convinced that he had seen the companion again, and his visit to England in May of that year was the occasion of his second (oral) communication to the Royal Astronomical Society.[2] He wrote a summary of this to Newcomb, giving measurements of the position of the companion that he and an assistant had made. The star's position had changed slightly and Otto found the change to be in agreement with a theoretical orbit calculated by the German astronomer Auwers. Otto promised to telegraph Newcomb on any night in the following spring that the companion was visible, so that Newcomb could look for it the same night. By April 1875, however, Otto had not had any nights on which he could be fully confident of his measures, and instead of telegraphing he wrote:

Until now I have, in fact, heard only of a single observation by Talmage, which confirms our observations here, and I confess that the circumstances that you still have not recognized the companion makes me uneasy in respect of my own observations [5]

By December of that year, Otto even asks if his senses are deceiving him, or "Are we dealing here with a body only occasionally luminous?"[6] He concluded that it would be unscientific to ignore his observations, confirmed by two assistants, not to mention Talmage's observation and the agreement with Auwers' orbital computations, and then he added:

But, to be sure, no-one longs more sincerely that I myself to see my observations confirmed by another telescope, if possible more powerful than ours [6]

Still Newcomb failed to confirm the companion and sent a paper he had written to Otto, who found that it only deepened the puzzle. All sorts of ideas now started up in Otto's mind:

Have we not here perhaps to do with phenomena such as we see in the Orion nebula, for which the appearance and disappearance of small points of light is an established fact? Is perhaps Procyon a star still in the process of development?[7]

These two sentences, paradoxically, show both how much astronomy has changed in the last hundred years and how, despite that, some underlying ideas have remained remarkably constant. We do not regard the appearance and disappearance of small points of light in the Orion nebula as "an established fact", but we do still believe it to be a region in which stars are being formed.

Similarly, we look upon the companion of Procyon as a star at the end of its evolution rather than the beginning, yet we agree with Otto that stellar evolution is important in understanding the system.   Otto may seem to be grasping at all sorts of unlikely straws in this last letter, yet he was trying to find a scientific explanation for his observation. Some readers may think that the possibility he raised himself in this same letter, that his eyesight was deteriorating with age (he was nearly 60) might have come nearest the mark. The truth, however, was more curious.   Otto withdrew his claim to have discovered the companion of Procyon (although he did not do so in the pages of *Monthly Notices*, where he first published it). Newcomb tells us the story in his own autobiography.[8] Despite Otto's conviction that the fault did not lie in the telescope, the objective of the 15-inch formed a ghost image of every bright star, and Otto and his assistants had taken this image for the companion of Procyon they had expected to see. Even the apparent changes in position could be explained by somewhat different positions of the telescope in the different years that the "companion" had been observed. Talmage had observed another star, near Procyon in the sky, that was already known.   Procyon's real companion was not discovered until 1896, when it was found by Schaeberle with a telescope even greater than Newcomb's --the 36-inch refractor of the Lick Observatory.[9] We now know that Otto's observations did not fit at all the well-determined modern orbit of the system.

Many of the same letters from Otto to Newcomb in which the companion of Procyon is discussed are dominated by descriptions of Otto's preparations for observing the transit of Venus in 1874.   An English astronomer, Jeremiah Horrocks, was in 1639 the first known person to observe such a transit. He had predicted himself that the event would occur on December 4 (November 24, Old Style) which was a Sunday. He could not predict the time accurately and had to sandwich his observing between church services. · (Horrocks is often described as a clergyman. This is not known for certain, but it does seem that he had to stand in for an absent cleric on this particular Sunday.[10]) He was able to observe part of the transit, despite some clouds. Later, Edmond Halley pointed out that transits of Venus provided an opportunity to measure the distance of the Sun from the Earth. Like Newcomb's work on the parallax of Mars, the method depends on the fact that a measurement of the distance between any two bodies in the solar system gives the scale of the system and enables all other distances within it to be deduced.   As seen from widely separated points on the Earth's surface, Venus will appear to cross the Sun's disc along very different paths, and from careful measurements of these, the distance of Venus from the Earth can, in principle, be determined.   The relationship between the orbits of Venus and the Earth causes transits to occur in pairs, eight years apart, but with each successive pair separated by more than a century. Halley, who died in 1741, never saw a transit, since the first pair after the one observed by Horrocks was in 1761 and 1769. The next pair was in 1874 and 1882. There have been no transits in the twentieth century so few people --if any-- alive today can have seen one; but the next pair will come conveniently early in the twenty-first century --in 2004 and 2012.

Because the phenomenon is rare and its observation seemed potentially so valuable, astronomers were prepared to go to great lengths to observe the eighteenth-century transits. Long and difficult journeys had to be made to

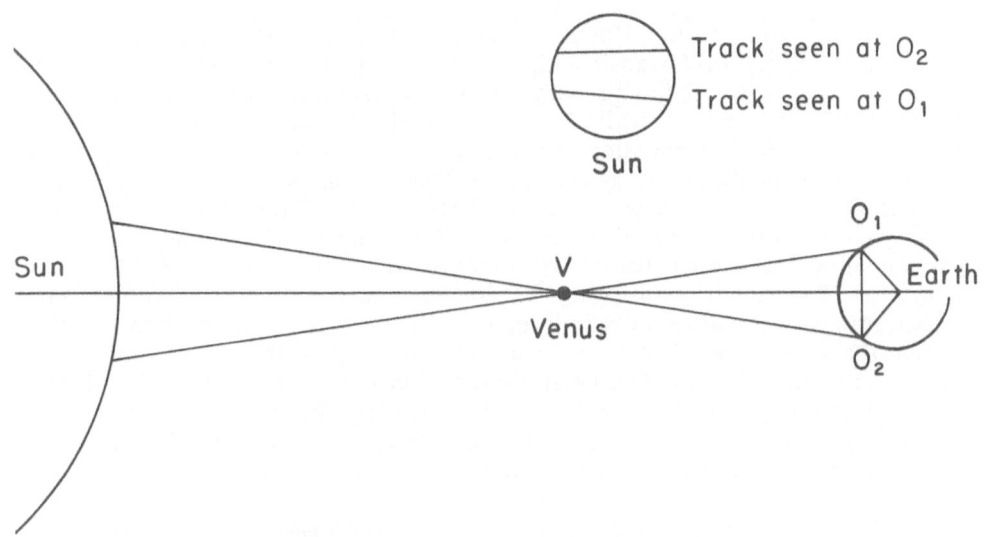

The principle of measuring the distance of the Sun from the Earth by means of a transit of Venus. Observers at 0_1 and 0_2 can, from their observations of the track of Venus across the Sun's disc (see upper sketch) deduce all three angles of the triangle $V0_10_2$. If they know the size of the Earth, they can compute the length 0_10_2 and hence the distance of the centre of the Earth from that of Venus. The distance from the Earth to the Sun follows from known properties of the solar system. To make the principle clear, the diagram is deliberately not to scale; in particular, Venus and the Earth are really almost the same size.

widely separated points in the hemisphere of the Earth that would be in daylight during the transit. Even European wars did not deter astronomers from these hazardous journeys, and the exploits of some were nothing short of heroic.[11] The 1769 transit was the occasion of one of Cook's famous voyages to the Pacific Ocean. The results were disappointing. In principle, an observer at a known position need determine only the instants at which the planet had just entered and was just about to leave the Sun's disc (the *second* and *third* contacts), and the positions on the disc where these contacts were observed. In practice, as we now know, this cannot be done accurately because of the thick atmosphere of Venus, which draws out the time of contact and gives the observer the impression that --for a while-- the planet is attached to the edge of the Sun's disc by a "black drop". By the late nineteenth century, the distance to the Sun was still not well known, and it seemed worthwhile to try again to observe the transits. No-one who had observed the eighteenth-century transits was still alive and astronomers had become sceptical about the reality of the black drop. Many hoped that the new method of photography would allow them to record the complete track of the planet across the Sun's disc, reducing the importance of any uncertainty in the measurements of the times of contacts. Another cause for optimism was that longitude differences between the different observing sites could now be measured by telegraphy, so the true distances between the sites were more accurately known.

Almost the entire Russian empire was favourably placed for the observation of the 1874 transit and, naturally, Otto masterminded the Russian arrangements. His old friend Airy likewise coordinated all observing in the far-flung British empire, while his new friend Newcomb, although a junior member of the U.S. commission charged with making American plans, took all the initiatives in the western·hemisphere (for reasons amusingly explained in his autobiography[12]). These three men between them, therefore, controlled a large part of the efforts to observe the 1874 transit, which dominated their correspondence for some years. As early as 1868, for example, Otto assured Airy that "Russia will do its duty"[13]. Once again, the support of Grand Duke Konstantin was enlisted and readily given. Since the Grand Duke had a special responsibility for the Navy, his support was useful in the setting up of observing stations on Russia's Pacific coast. Perhaps because Otto visited Airy fairly frequently in the years before the transit, more details of the arrangements are found in the letters to Newcomb. The first in which the topic was mentioned was written (in English) in March 1872:

Last summer I sen[t] to the Naval Observatory copies of the protocols of the Russian Venus-commissions. Though they were written in Russian, I hope you will have been able to find at Washington a translator. In addition I beg leave to send now a copy of a letter I sent a fortnight ago to Mr. Airy:

"The inquiries on the meteorological conditions of the selected stations have given on the whole very satisfactory results, particularly for the stations on the coast of the Pacific Ocean and in Eastern Siberia (85% clear sky for December). There are only two of the selected stations, Tashkent and Astrabad, for which these conditions are not sufficiently satisfactory. Therefore the observers designed for Tashkent will probably go to a place about 100 miles west of that town, and instead of Astrabad we shall either take the island of Aschuradeh in the Caspian Sea or, if possible, cross the Elburz mountains and establish our observers at Schahrud in Persia (with near absolute certainty of clear sky).

"The total number of Russian stations will be 24, each of them provided with <u>only one</u> instrument for the Transit observation. [Here follows a list.] Each station will be furnished besides with clocks, chronometers and the instruments necessary for the determination of time.

"...For these instruments the astronomers are also in a great part allready [sic] committed. They will come to Pulkowa for a certain time next year, to exercise themselves in the observations.

"The geographic positions of the stations will not be determined by Transit observers, but all stations, on which the Transit has been successfully observed, will be carefully determined afterwards by special expeditions of the General Staff or Navy. For this purpose a principal line of telegraphic longitudes will be laid, probably next year, through all Siberia to Nikolayevsk, with which line the other stations of that part of Russia will be easily joined either by telegraph or chronometric operations.

"With regard to the photographic method, I can inform you that in two places: at Vilna (under the direction of Mr. K[norre?] and at Bothkamp in Holstein (Dr. Vogel), it has perfectly succeeded to take instantaneous photographs of the Sun with dry plates"[14].

The transit was not visible from the United States, so Newcomb needed this information to make his own plans. While one American expedition was to the island of Kerguelen, in the southern hemisphere, where American and British teams observed (or, rather, were clouded out) side by side, another was planned to the Pacific coast of Siberia. Otto welcomed the proposal, subject to the Tsar's approval, and he wrote to Newcomb about it early in 1873:

Let us chat about the preparations for the observation of the transit of Venus, which just now occupies me much. It goes without saying that we astronomers would be very happy if the American commission were to occupy also a station in the neighbourhood of ours, in the Russian coastal region on the Pacific Ocean. Thus would be provided particularly a very useful control for the various methods of heliographic determination carried out by both sides. Also our government is quite in agreement with it and the Grand Duke Constantin, with whom I have spoken about it, has authorized me to propose that the American side should choose the capital of this coastal region Wladiwostok as the station. This place ($\varphi = 43° 6'7$, $\lambda = 8^h 47^m 45^s$E. of Gr[eenwich] was originally designated as a Russian station; we can, however, just as well establish our observer on a neighbouring point, since he is familiar with the local language, roughly 10 Engl. miles away on the island of Nachodka. What makes Wladiwostok especially advisable as an American station, apart from the geometrically [presumably Otto meant geographically] advantageous position and the very suitable meteorological conditions, is the circumstance that there, in the capital of the coastal region, more resources can be placed at [your] disposal, than in any other place, and that during the winter the harbour is closed by ice only for a very short time-- will not be compelled in an unnecessary way to a long delay there.[15]

Otto goes on at some length about the technical details of the various instruments he proposed to use. His letters to Newcomb, often eight or nine pages long, were obviously written in a hurry: the handwriting became crabbed and the syntax often suggests that his mind was racing ahead of his pen, so that his precise meaning is sometimes hard to decipher. He was aware of these faults and apologized for them, usually ascribing them to pressure of other correspondence or the "unaccountable" quantities of paper-work with which he had to deal. In one letter he offered another excuse:

I was many times interrupted in the writing of this letter and must on that account ask indulgence for some carelessness. In fact, I have in this life a busier time than I have ever known before. That is principally Venus' fault.[16]

Even so, in most of his letters to Newcomb, Otto found time for a little gossip:

At the Paris Observatory, complete anarchy has reigned for three months and even now the situation does not appear to be appreciably more agreeable. Under Delaunay's direction it got even [worse] than under Le Verrier. He was a scholarly and good-natured man, who, however, did not understand administration. Therefore it comes about that the subordinates are not only constantly opposed to the Director, but also quarrel among themselves and disturb each other's work. It is a phenomenon worthy of note, and also explicable only by the French character, that the same men who for three years pressed for Le Verrier's dismissal, are now inclined to petition for his reinstatement in the directorship. They have now learned that a strong hand over them is needed for them to be able to work in peace and quiet. I believe and hope that Le Verrier will come back to the helm.[17]

Le Verrier did come back, but not for long. He died in 1877: "A great mind is departed with him!"[18] wrote Otto. Newcomb had had difficulties with Le Verrier and feared the latter's reinstatement as Director of the Paris Observatory might adversely affect his own attempts to collect and examine old observations of Jupiter's satellites. Otto maintained that he had

...always found Le Verrier very ready to support scientific missions, and because of his great energy this support is always powerful. But frankly he is not a character easily handled, and therefore I may boast that during 30 years I have sustained friendly relations with him.[19]

By all accounts, Otto was justified in "boasting" of this achievement. Perhaps he was helped by the strong mutual respect that had existed between his father and Le Verrier. Otto's letters often display a certain prejudice against the French, as does one of the above extracts. Nevertheless, he remained on friendly terms with a number of individual French astronomers.

By June of 1873, Otto had learned that Newcomb had accepted the proposal of Vladivostok as an American observing site. Later that summer, Newcomb came to Europe and he and Otto met in both Hamburg and Hanover. Otto's own travel plans were rapidly revised when he learned of a cholera outbreak in Le Havre. He cut out an intended visit to France, and went to the Swiss side of Lake Geneva instead, calling at Strasbourg, where he discussed the German plans for observation of the transit with Winnecke and Auwers. Back in Pulkovo, in January 1874, Otto wrote to Newcomb that he was "in general well satisfied with the progress of our preparations",[19] although he was worried about the lack of progress in the proposed series of telegraphic longitude determinations, which depended on European cooperation for full success. He found that "Our German friends are too pedantic --or tactless-- for that kind of negotiation".[20] Diplomatic problems raised by the American plans to observe from Vladivostok had all been straightened out, and Otto promised that naval-lieutenant Onazawitsch "who speaks good English" would be on hand to help Professor Hall and his companions. In a later letter Otto was concerned how Hall would get from Japan to Vladivostok and suggested that he might like to make that part of the journey in a ship of the Russian Navy.[24] In fact, Hall travelled all the way in a ship of the U.S. Navy that waited in Vladivostok until he was ready to return. It is hard to know which of these courses would be more agreeable to both parties now!

In May of the transit year, Otto travelled to western Europe again. Although he found the English [sic] Admiralty ready to help in the telegraphic longitude determination, other nations were not, and he regretfully decided that Russia must rely on her own resources in this aspect of the work. He learned that Airy was planning an expedition to Hawaii (then still a British protectorate known as the Sandwich Islands) and was anxious for American help in determining the longitude differences between San Francisco, Hawaii and Yokohama. Otto also visited France to learn about the French plans. He wrote about all this to Newcomb in a letter from Greenwich,[21] which seems to have been the last that he wrote before the transit. Airy was favoured with a long letter written on the night before that event, but Newcomb had to wait until afterwards --two weeks afterwards, if the letter is dated on the Gregorian calendar-- to learn, as astronomers so often do, that the carefully made plans had been largely frustrated by factors beyond their control.

> That Russia was very little favoured by the weather for the transit of Venus will have been announced to you by the newspapers. Of our 31 stations, 15 have had completely covered skies, 8 only partly clear, 4 quite clear and from 4 reports are still lacking. Despite this, however, the observations here, to be sure, will provide a valuable contribution to the parallax determination; in particular, I expect that from the heliometer measures obtained in Nertscheurh and Tschita --on the significance of which so much the more weight must fall, if the Egyptian telegraphically announced report "that a considerable Venus atmosphere was observed" is confirmed. As it seems to me, such an atmosphere, namely, may essentially modify the contact observations while the distance measured on the Sun will not be influenced by it. Naturally, the photographic exposures obtained by the Sun would also have the same advantage.

> That your observers in Wladiwostok also have had to share the unsuitability of the weather with ours, makes me sincerely sorry. I have been told, however, that they may have obtained 10 good photographic exposures and even that is certainly valuable. Whether your observers in Wladiwostok found the reception and support on which they might with right count is a matter in which I have some worrisome care, namely that the Governor of this region, who had received in this respect the necessary orders to this effect, as first learned later, perhaps did not return to his post and also in other respects appears to have pursued the matter very indifferently.[22]

Otto's family had been involved in the expeditions. Döllen led the party to Egypt --which did no good to his health-- and Hermann was sent to observe in Siberia. He returned through Japan, visiting his Uncle Karl who was Russian Ambassador in Tokyo. Most participants, in all countries involved, felt that the effort put into the preparations had produced only disappointing results. This attitude was summed up by Otto when he wrote to Newcomb some years later about the second transit:

> That Russia does not have the intention officially [Otto's emphasis] to take part in the observation of the 1882 transit of Venus you will perhaps already have heard from other quarters. We are happy to leave this observation to others who hold the view that in this way a more exact value of the parallax can be obtained, than continual opposition observations of Mars and the minor planets, from one place, east and west, can deliver. However, apart from the, in my opinion, fruitless trouble to obtain the most exact parallax determination from the transit, the phenomenon itself remains a strongly interesting [one], and I would willingly support it, in case by any chance a Russian astronomer should express a special wish for the purpose to transfer to any suitable station situated in America.[23]

Sir David Gill   (Photograph reproduced with the permission of the Royal Astronomical Society)

The 1882 transit was visible over the whole of the United States. Otto, by that time, was increasingly preoccupied with the construction of a new large telescope for Pulkovo, the lens for which was being ground in Massachusetts (see next Chapter). He had in mind that a Russian astronomer might visit America to conduct the acceptance tests on the lens and to observe the transit. Events did not work out that way. Newcomb makes clear in his autobiography that he personally agreed with Otto's assessment of the value of transit observations for the determination of the solar parallax.[24] He wished to confine American efforts to observation in the U.S. The majority of American astronomers wanted another global effort, however, and the government provided more money than anyone expected. With Otto's further help, a second expedition was sent to Siberia and Newcomb took the opportunity to lead an expedition to the Cape of Good Hope, where he was welcomed by Gill. The results were again disappointing and, according to Newcomb, never fully published. Now that distances within the solar system can be measured extremely accurately by radar, it is most unlikely that the next pair of transits will be the occasion of such large-scale cooperation between nations.

Airy, the third principal actor in 1874, was now removed from the scene. After nearly fifty years on the job, he resigned from the post of Astronomer Royal in 1881. In September of that year, Otto for the first time mentioned his own retirement to Newcomb, and freely expressed his doubts about Airy's successor:

> These thoughts have been particularly lively in me since the report of Airy's retirement from the office of Astronomer Royal arrived. I find his decision completely justified, even if it must, perhaps, be said that a substitute for him in Greenwich will be even harder to find than would be the case in the somewhat analogous case for me here in Pulkowa. Whether and how far Christie will grow in the task must be taught by experience. He is still, indeed, to some extent a homo novus, of whose capacity all who are not personally acquainted with him can scarcely well form a judgment on only his existing works. But just because he is still an unknown quantity, the hope of a solid and skilful performance remains not excluded, while the other competitors of whom there was talk (Stone, Proctor, Lockyer) from here indeed seem to be very questionable. It will please you, I think, to hear that with the first report of this retirement, here in Pulkowa, quite independently on many sides was expressed "What a pity Newcomb is no Englishman".[25]

Between the transits, Otto got to know David Gill, who came to astronomy late in life, but was appointed H.M. Astronomer at the Cape of Good Hope in 1879. In preparation for his work in this new post, Gill made a tour of major European observatories, including Pulkovo. Otto wrote to Newcomb "...the exchange of Gill for Stone is a significant gain for exact observations in the southern hemisphere".[25] Otto's supreme compliment to Gill, comparing him to Wilhelm, has already been quoted (Chapter 4). Kapteyn made the same comparison when he wrote Gill's obituary.[27] Gill was delighted with Otto's compliment:

> Your father has always been my beau ideal of what a great practical astronomer should be, and no compliment could be higher than to compare my work in a remote way with his -- alas, how far behind no one knows more fully than myself. In meridian astronomy in the present day all the world except Pulkowa is behind the equipment and system which he

Sir WHM Christie eighth Astronomer Royal of England   (Reproduced by permission from the collection of presidential portraits of the Royal Astronomical Society)

instituted about 1843 [date barely legible and possibly uncertain] and which has been carried on by Pulkowa ever since [28]

These letters were exchanged in 1886, after some years of friendship, but even in 1881 Otto asked his new friend how Airy's retirement would affect *him*, and was even more openly critical about the new Astronomer Royal than he had been to Newcomb·

[Christie] never has made upon me the impression of being a very superior man and until now no standard work has been issued by him   Let us hope there is more in him as he has shown until now [29]

Gill, however, sprang to the defence:

> Christie's appointment pleases me much. We have always been good friends together, and
> there is much more in the man than has yet come out. I am perfectly happy here , I have
> a grand [?] field for work before me, and I would not leave the Southern Hemisphere
> without doing a good piece of work in it. I have cut out a programme to last me for 15
> years at least, and I care more for my programme than for any position.[30]

Thus admonished, Otto tried to like Christie but, as we shall see, sometimes found it rather difficult.

As the friendship between Otto and Newcomb grew, letters to the latter as well as those to Airy, give frequent glimpses of Otto's family life. Even the long and somewhat disheartened letter about the unsuccessful transit observations mentions Otto's lively little daughter (*kleiner Töchterchen*) Eva, and that nine months ago he was "through his second son Alfred [m. 1872] promoted to grandfather". Surprisingly, Otto was only six years behind Airy in this "promotion". Newcomb invited Otto to visit the United States in 1876 to take part in the celebration of that country's centennial. Otto declined because he felt that the Russian government would not approve of its own man of science making a speech on a political occasion.[31] His opportunity to go to America came a few years later.

The following year, after referring to the forthcoming marriage of his daughter Thérèse, Otto mentioned his son Hermann for the first time to Newcomb. Hermann, after his return from Siberia and Japan, had "completed his university studies as a mathematician and shows much skill".[32] Otto planned to send him abroad for further study, perhaps to America. Like his father, Hermann had to wait for that opportunity, but he did study with the astronomer Winnecke, the physicists Kirchhoff and Helmholtz, and the mathematician Weierstrass.

Two years or more before the first transit of Venus, Otto's old friend Airy received an honour that called for congratulation. Although, chronologically, it belongs in the beginning of this chapter, the somewhat naughty letter that Otto wrote for the occasion provides a good note on which to conclude:

> Dear Sir,
>
> Having just read in the newspapers that you have been promoted to the Knighthood of
> the Bath I beg leave to express you that it had given me great pleasure to hear of this well
> earned distinction. --However I have been somewhat puzzled by apprehending that you
> have now accepted that distinction, which you have told me, you have refused several
> times. If 30 years ago you had been made a Knight, your friends would have been much
> more pleased than now. What can Mr. Airy gain, by being titled "Sir"? What
> circumstances have acted to change your former views? --If you have not the time to give
> me the key to the enigma, perhaps Christabel [Airy's daughter] will do it.
>
> Please remember me kindly to Lady Airy and your children. I hope you are all well.
>
> Ever truly yours   Otto Struve.[33]

Airy in retirement, in the garden of the White House, Greenwich, with his surviving family. Back row: Wilfrid, Hubert, Annot, Christabel, Osmund. Front row: Anna (Wilfrid's daughter), Emily (Osmund's wife), Airy and Hilda. (Reproduced by permission of the Royal Greenwich Observatory, Herstmonceux.)

Such is Airy's posthumous reputation that many might imagine that he took grave offence at this letter. Either our judgment is mistaken, or Airy and Otto were very close friends --perhaps both are partly true-- for, in the opening paragraphs of his reply Airy rose to the occasion with a (somewhat mathematical) joke and then gave the sort of discourse on the English orders of chivalry that, many years earlier, he had solicited from Otto's father on the Russian:

My Dear Sir,

Your letter of July 1 has been lying before me a long time, but I have had successions of business coming on for some time faster than I could clear them off. But lately the flow has greatly eased, and the effort of overcoming them predominates (mathematically speaking, the differential coefficient has changed its sign), and I am finding leisure to answer letters.

First, then, about the rank which in English is known by the three letters K.C.B. This is an honor [sic!] very much valued. There are above it the English Order of the Knights of the Garter, the Scotch Order of the Knights of the Thistle, and the Irish Order of the Knights of St. Patrick. All these are for Noblemen only. Next is the order of the Bath. The division G.C.B. is much limited: and K.C.B. follows it. I think I can truly say that I care very little for such things: but the public estimation of this (wondrously different from simple knighthood) is well known: the Prince of Wales warmly congratulated me, and my family naturally are pleased with it.[34]

NOTES

1.   O.W. Struve, 1873, Monthly Notices of the Royal Astronomical Society, Vol. 33, pp. 430-33.

2.   O.W. Struve, 1874, ibid., Vol. 34, pp. 335-9.

3.   O.W. Struve, 1873, letter of June 3 to Simon Newcomb.

4.   O.W. Struve, 1874, letter of January 21 to Simon Newcomb.

5.   O.W. Struve, 1875, letter of April 12 to Simon Newcomb.

6.   O.W. Struve, 1875, letter of December 11 to Simon Newcomb. The confirming observation is ascribed to Tebbutt in this letter, but Otto corrected himself in a later letter.

7.   O.W. Struve, 1876, letter of February 11 to Simon Newcomb.

8.   Simon Newcomb, Reminiscences of an Astronomer, (see Chap. 6, ref. 30), pp. 138-40).

9.   J.M. Scaeberle, 1896, Astronomical Journal, Vol. 17, p. 37.

10.   W.T. Bulpit, 1914, The Observatory, vol. 37, pp. 335-7.

11.   J.D. Fernie, 1976, The Whisper and the Vision: Voyages of the Astronomers, Clarke Irwin and Co., Toronto, Chap.1.

12.   Simon Newcomb, Reminiscences of an Astronomer, (see Chap. 6, ref. 30), pp. 160-2).

13.   O.W. Struve, 1868, letter of November 10 to G.B. Airy. (RGO 951 f. 759).

14.   O.W. Struve, 1872, letter of March 16 to Simon Newcomb.

15.   O.W. Struve, 1873, letter of February 4 to Simon Newcomb.

16.   O.W. Struve, 1873, see ref. 3.

17.   O.W. Struve, 1873, see ref. 15.

18.   O.W. Struve, 1877, letter of December 7 to Simon Newcomb.

19.   O.W. Struve, 1873, see ref. 3.

20.   O.W. Struve, 1874, see ref. 4.

21.   O.W. Struve, 1874, letter of May 9 to Simon Newcomb.

22.   O.W. Struve, 1874, letter of December 23 to Simon Newcomb.

23.   O.W. Struve, 1881, letter of September 1 to Simon Newcomb.

24.   Simon Newcomb, Reminiscences of an Astronomer, (see Chap. 6, ref. 30), pp. 173-4.

25.   O.W. Struve, 1881, see ref. 23.

26   O.W. Struve, 1879, letter of March 30 to Simon Newcomb.

27.   J.C. Kapteyn, 1914, Astrophysical Journal, Vol. 40, pp. 161-172.

28.   D. Gill, 1886, letter of September 14 to O.W. Struve. (RGO 15/126 ff. 887^r to 892^r).

29.   O.W. Struve, 1881, letter of August 18 to D. Gill (RGO 15/125 ff. 549^r to 550^r).

30.   D. Gill, 1881, letter of September 19 to O.W. Struve. (RGO 15/125 ff. 551^r to 554^r).

31.   O.W. Struve, 1876, see ref. 7.

32.   O.W. Struve, 1877, see ref. 18.

33.   O.W. Struve, 1872, letter of July 1 to G.B. Airy, (RGO 952 f. 662).

34.   G.B. Airy, 1872, letter of July 23 to O.W. Struve. (RGO 952 f. 623).

CHAPTER 14

THE 30-INCH REFRACTOR

After Otto's death, his friend and long-time colleague Magnus Nyrén wrote of him to the effect that he had taken his father's creation (i.e. Pulkovo) and raised it to a double height.[1] In one respect this was literally true: Wilhelm had endowed Pulkovo with a 15-inch refractor, Otto provided it with one of 30 inches aperture. At the time of the celebration of Pulkovo's fiftieth anniversary, Otto wrote an official description of the acquisition of this large telescope.[2] His letters to Newcomb, Gill and Abbe provide a less formal account and allow us to follow the development of the project with the many difficulties that Otto encountered at the time. Newcomb's influence was particularly important --he himself claims to have suggested that the firm of Alvan Clark and Sons might grind and polish the great lens-- and Otto made clear from the beginning that Newcomb's opinion carried great weight with him. Otto first broached the matter to Newcomb in a letter written in March 1879. He explained that it had been Wilhelm's deliberate policy to provide Pulkovo with the most powerful refractor in the world --just as earlier he had done for Dorpat. The original 15-inch had been equalled by Harvard, however, and now surpassed by Washington. Indeed, an equal to the Washington telescope was even then under construction in Vienna. It troubled Otto that Pulkovo should thus be dethroned from the position his father had envisaged for it, but he made no approach to the government until after the end of the Russo-Turkish war of 1877-8, when he proposed the construction of a refractor of 30 to 32 inches aperture for Pulkovo because

> ...according to an investigation by Mr. Simon Newcomb these dimensions form nearly the limit beyond which not considerably more for "space penetrating power" [English phrase in a German letter] can be obtained. Of this your speculation, unfortunately nothing else is known to me than what is said in a notice of your Popular Astronomy in the Monthly Notices (I have not yet set eyes on the book itself). It would interest me much, therefore, to learn whether even now you are still of this opinion.[4]

Newcomb was not far wrong: refractors of up to 40 inches (1 metre) were built before astronomers began to understand the superiority of reflectors at such large apertures. Otto's proposal had been supported by several influential people (he did not name them) and he had been instructed to investigate whether or not such an instrument could be successfully built in Europe. Accordingly, he had spent the autumn of 1878 visiting Vienna, Munich, Paris, London, Birmingham and Hamburg and concluded that the glass blanks for the lenses should come either from Chance Brothers or Merz, that the latter should grind and polish the lenses and that Repsolds' --who had built so much for Pulkovo and for whom Otto had worked for a while in his youth-- should build the mounting. Otto even had chosen the site for the new building needed to house the telescope, but he hesitated to accept the bid from Merz, partly because he thought their price too high

> ...more, however, because of a certain indecision and vagueness that manifests itself in all dealings with Merz.[4]

This led him to ask Newcomb about the firm of Alvan Clark and Sons, in Cambridgeport, Massachusetts, who had built the Washington telescope and thereby demonstrated their ability. (Otto's letters do not bear out Newcomb's memory that he first made the suggestion.) Otto did not want to make a trip across the ocean, however, and proposed first to make further inquiries in Munich in May or June.

> In the meantime, however, it would only be very desirable to obtain from you a candid expression of opinion about Messrs A. Clark. The old Mr. Alvan Clark must now already be quite aged, so that, to be sure, it can scarcely be counted on that he himself will carry out the work. Are his sons just as skilled and reliable as the father was? That is the principal point.[4]

Otto was perhaps the more inclined to look to the U.S. since the construction of the Vienna telescope (undertaken by European firms) was encountering several difficulties. He was also impressed by the focal ratio of the Washington telescope, f/16. Merz would grind the lenses to the traditional Fraunhofer ratio of f/18, but Otto recognized that the shorter focal length (and, therefore, physical length) of the instrument would enable considerable savings to be realized in the cost of the dome.

Newcomb at once started making enquiries and answered Otto's letter promptly and persuasively. Early in May 1879, Otto wrote again:

> 1. I am firmly resolved, if no special obstacle intervenes, to visit you in America in the coming summer, in order to familiarize myself with what has been already achieved in the optical area and also to form my own opinion on what is to be expected.

> 2. I am very disposed to break off completely the negotiations with Merz (especially since the latest correspondence has allowed his indecisiveness to come forward almost blatantly) and to place the order with Messrs A. Clark and Sons, but I reserve my decision until my visit and the personal negotiations.[5]

As it happened, Otto himself proved somewhat dilatory in making the visit. At the same time as these negotiations for a new big telescope were being made, the 15-inch was to be provided with a new mounting by the Repsolds, and the old dome was being modified and repaired. Otto had to stay to supervise this work, and his first stop on his journey abroad was to be in Mitau, for the marriage of his oldest son Georg Wilhelm on June 29th. He then planned to visit Repsolds' in Hamburg and the Chance Brothers in England. He hoped to arrive in Washington by August 1st, but in the middle of June he made his first postponement:

> In 5 days I leave here and if no illness or something else quite extraordinary occurs, I will be with you in Washington around the 10th August. My son Hermann, who is coming

Hermann Struve in 1877. The back of the original of this photograph (by Alfred Lorens, St. Petersburg) is inscribed by Hermann to his cousin Arvid Lindhagen.

along nicely as a competent young astronomer, or rather mathematician, will accompany me on the journey; I am already too old to undertake one so far without [a] companion, and, moreover, I expect that this journey will be quite especially instructive for him.[6]

On each step of the journey, however, Otto and Hermann got a little more behind schedule, but they finally managed to book a passage on the *Scythia* of the Cunard line, and arrived in the U.S. around the 12th August. Newcomb, who no doubt knew what Washington could be like in August had gone north for a holiday in Massachusetts and, later in the summer, had to travel to California in connection with the building of the Lick Observatory. Otto wanted to observe with the Washington refractor and judge it for himself before meeting the Clarks. He strayed into a kind of summer he had never before experienced and had very little clear sky. His account, written to Newcomb *from* Washington, will immediately engage the sympathy of every observational astronomer (Newcomb, after all, was primarily a theoretician!):

This is now the 5th day that I am spending here in expectation that it will be possible for me to make some observations with the great equatorial. The weather, however, is enough to drive one to despair; always covered sky, almost continuous rain, and with a heat that quite melts away a northerner. Only on the first evening after our arrival could I convince myself that there are stars even in Washington; however, scarcely had I rushed out to the Observatory, than the sky was covered again and since then I have seen no stars again. This morning it also rains, but there seems to be a change taking place and a somewhat fresher wind is blowing from the north-west and that arouses again the hope that today or tomorrow I will obtain observations. --So soon as that happens I will hasten "by first through train" [English phrase in German letter] to Boston and hope accordingly to see you in a few days. In any case, I will announce telegraphically the time of my arrival to Prof. Pickering as soon as I can certainly, determine it.

In Boston resp[ectively] Cambridge then we will speak further. In the journey to San Francisco, I may not easily participate, tempting as the opportunity is. On the contrary, my intention is still firm if possible, to embark again, from New York for Europe, on the Gallia on September 10th.[7]

Otto and Hermann did reach Cambridge and met Newcomb there, as he recounts in his autobiography.[8]   The interviews with the Clarks were satisfactory and the firm was commissioned to make the new objective. While in the United States, Otto met many American astronomers, particularly, of course, those on the staff of the Naval Observatory. Naturally, he renewed his acquaintance with Cleveland Abbe.   (Later, in New York, just before embarking homeward, he met Abbe's mother and sister, and particularly appreciated their parting gift of a "little basket of splendid grapes and peaches"[9]). In Cambridge, Otto and Hermann met Pickering, the Director of Harvard Observatory. For a short break, they visited Niagara Falls, and after a day there they returned through Cincinnati to New York, but first they rested

... for a day in Clinton [N.Y.] with Peters, who had accompanied us to Niagara, and I was pleased about this, in that it gave me an opportunity to take a closer look at his admirable star-chart work. The painstaking care with which he executes and continuously improves these charts, certainly deserves the greatest recognition; his numerous discoveries of minor planets proceed, in fact, only from this, that each small point of light that he can definitely recognize and that is not plotted in the completed charts is a suspect object, that he can assume with the greatest confidence is either such a small planet or a faint variable star. Such carefully worked charts, however, are offered a still much wider and more important field of application in that they contain certainly the best and most trustworthy material for studies of the structure of our star-system. On this ground it is certainly in the highest degree desirable that not only the charts made already should be reproduced by printing or photography and be made generally available, but that care should be taken that the work should be brought to a conclusion, at least within the hitherto adopted limits, but if possible extended in a similar manner beyond the whole sky visible from Clinton. Although, as Peters said to me, the means for publication are not hard to find, still it must be well considered that he is now, if I am not mistaken, 65 years old and therefore, according to human measure, will not be able to keep on working with the same energy for many more years. It is, therefore, much to be recommended that a younger, more zealous astronomer should join him for the purpose mentioned. Peters himself would not only have no objection, but wishes it most fervently, but I doubt that, on account of his modesty and his characteristic faults from lifelong habit, he himself will take the steps perhaps necessary for it. For this reason his friends are called in the interests of science to take action instead, and I allow myself on this account to urge this matter on you [Newcomb]. Your experience and knowledge of the situation will easily let you discern which road should be followed in America for this goal. If the work were being carried out in Russia, I would know what I had to do; truly, not a month would go by without the necessary assistant being provided.[10]

There were three German-born nineteenth century astronomers called Peters. To add to the confusion, each had three initials, the first of which was C., and another F. One of the original four astronomers at Pulkovo was C.A.F. Peters. He returned to his native Germany and succeeded Schumacher as Editor of the *Astronomische Nachrichten*. His son, C.F.W. Peters, was also an astronomer. In the above letter, Otto wrote of C.H.F. Peters, apparently not connected with the other two. As a young man, this Peters had worked in Italy, but left that country (and later Europe) after the 1848 revolutions. He frequently returned to Europe for meetings of the *Astronomische Gesellschaft*, but remained a somewhat controversial character.

One American astronomer whom Otto would have liked to meet was the double-star observer S.W. Burnham . Before he left Russia, Otto wrote to try to arrange a meeting.[11] Burnham lived in Chicago, however, and as he planned to accompany Newcomb on the journey to San Francisco, he did not have the opportunity to meet Otto. Even three years earlier, Newcomb had written of Burnham's plans to prepare a general catalogue of double stars --a work which was not printed before Otto's death. Surviving letters give the impression that Otto would have liked to undertake this task himself but did not have the time --he was still working on the publication of his own double-star observations. He obviously had a high opinion of Burnham --at that time still an amateur astronomer, but famous for his ability to resolve pairs of stars that observers with much larger telescopes than he had, found very difficult-- and stood ready to give him as much help as he could in the compilation of the catalogue.

In America, for the first time, Otto got to know Newcomb's family. He was delighted to learn that Mrs. Newcomb was a granddaughter of Hassler, founder of the U.S. Coast and Geodetic Survey, who, before he went to the U.S., had been a friend of Martin Bartels. Of the Newcombs' daughters, Anita seems to have attracted both Otto and his half-brother Karl, who a few years later became the Russian Ambassador to Washington. Both men would conclude their letters to Newcomb with formal greetings to Mrs. Newcomb and the children, frequently adding a less formal message for Anita. She was only 15 in that summer of 1879, but already she seems not to have been afraid to tease Otto, and he seems to have enjoyed it --perhaps he sometimes got tired of the awe and respect that were the usual lot of *Herr Staatsrath und Direktor*. Anita was certainly an exceptional person; she studied in Cambridge (England) and Geneva and became a specialist in genealogy. She married W.J. McGee, a geologist and ethnologist, and --partly in order to help him --studied medicine and qualified as a physician. She organized the army nursing corps during the Spanish-American War and became the first woman entitled to wear the uniform of a U.S. Army officer.[12] Otto had a soft spot for her during the rest of his life.

The result of Otto's negotiations in America is expressed succinctly in the letter he wrote to Cleveland Abbe shortly after his return:

That I have ordered an objective for a 30-inch from Clark's you will have learned through the newspapers and in other ways. Alvan Clark Jr travelled with us to Europe to order the glass blanks and, as I hear, he has ordered the Crown glass from Chance and the flint from Feil in Paris. As is to be expected, the Clarks are setting about it with energy, and, therefore, the expectation is justified that they will succeed in supplying the finished objective in 2 years. If this expectation is fulfilled, I have, perhaps rather rashly, promised that I myself will come over yet again to test the objective and accept it. As much as the promise of seeing my American friends again will please me, yet I must describe this promise as rash, however, since this year's journey shows me only too well that I am no longer young and must not impute too much to my strength.[13]

Otto complained to both Newcomb and Abbe that he had felt extraordinarily weary all the time that he was in the U.S. To Newcomb, indeed, he wrote that he must have "seemed like a very weak old man".[14] On his last night in New York, Otto had a violent fever for several hours, which cleared up, but then recurred both on the ship and in England. By the time he reached Germany, he had developed a heavy cold and remained ill until after he reached home. He remained very much in two minds about making a second trip to America, sometimes declaring that he could not do it and at other times entertaining the idea of accompanying Newcomb to California. He was much interested in the plans for the Lick Observatory and frequently asked Newcomb about their progress.

As might have been expected, the 30-inch objective --the largest yet to have been attempted-- encountered a number of problems and delays. Chance Brothers had an accident with the crown-glass component in the winter of 1879-80 and Feil then undertook to supply both the components needed for the achromatic objective. He had problems too, delaying Otto's negotiation with Repsold over the cell for the objective. At the same time, Otto had family worries: his son Alfred, now the father of four children, had a severe attack of typhus, that same winter. Alfred recovered, but his sister Thérèse died in childbirth shortly afterwards. By August, Otto was free to worry about the telescope again. He had heard no word from either Chance or Feil, and doubted (with reason) that the objective would be ready for testing in 1881, as planned. The new mounting for the 15-inch was expected in September 1880 and plans for the new dome for the 30-inch were advancing, despite struggles with the architect. In November, Otto heard that the Clarks were well pleased with the crown-glass and hoped soon to have the flint, but not until August of 1881 did he have more definite news (which implies that more problems had been encountered with the crown-glass) that he wrote about to Gill:

Last week I have been informed that at last, after several mishaps, the moulding of the crown-glass lens for our 30-inch objectif [sic] has perfectly succeeded (the flint is allready since a year in the hands of the Clarks). Acting upon this information I intend going within 3 weeks to Hamburg for contracting with the Repsolds about the mounting and discussing with them all particulars of the construction. This being done it will be time to go to Strasbourg for the Astronomische Gesellschaft meeting, which this year from different circumstances will probably be more important than usually. After the meeting I shall probably extend my journey to Italy and the South of France. My son Ludwig, of whom I shall send you in the next few days a paper on the irregular proper motion of η Cassiopeiae, will accompany me on this journey.[15]

This is the first mention of Otto's youngest son, Ludwig, now 23 years old. He had studied at Dorpat, and this journey appears to have been his introduction to astronomers outside Russia. Otto also wrote about Ludwig to Newcomb, commenting:

> As it appears, he will turn completely to astronomy, while Hermann has made a turn towards mathematical-physical theory.[16]

In that same letter, Otto explains more about the purposes of the journey. Schiaparelli was also ordering a large telescope for Milan, and was to meet Otto and Ludwig at the Repsolds' factory. Together they would go to the meeting in Strasbourg (Winnecke would be the host: it was the last year before his second breakdown) and then to Milan, where they would prepare for publication the "numerous and so excellent double-star measures of Dembowski".[16] Baron Ercole Dembowski was another amateur astronomer whose observations of double stars have, by their number and precision, assumed a classical importance.

From Milan, Otto was to go to Nice, where the banker R.L. Bishoffscheim had endowed an observatory that was to be equipped with a large refractor. Bishoffscheim proposed to order the objective from the Clarks and wanted Otto's advice. After his visit, Otto expressed to Newcomb some doubts about whether the site for the new observatory had been well chosen:

> The position of the Observatory, 1500 feet above Nice on an isolated hill, is in [a] picturesque respect incomparably beautiful, in that to the south the Mediterranean Sea stretches as far as one can see, while to the north the horizon is limited by the snow-peaks of the Alps. Whether, however, in astronomical respects the site chosen is appropriate may well be very questionable, since I very much fear that, at least with certain wind directions, the contrast between the Alps and the Sea could produce great instability in the seeing, although at other times, to be sure, a very transparent sky would rejoice those there. The future must teach, but it is still a pity that no trial was made, in this respect, with sufficiently powerful instruments. To be Director of the Observatory, that anyway will be quite 3 or 4 years first coming into full operation, M. Perrotin is appointed, who has gone to a great deal of trouble, even outside France, to prepare himself for the task before him. One-and-a-half years ago he visited us here in Pulkowa and we had a real liking for him...[17]

Otto's prime concern, however, was still his own 30-inch. In the letter just quoted he mentioned that he was still negotiating the focal length with the Clarks, but he hoped that the final tests could be made in the autumn of 1882. Now that Karl had been posted to Washington, Otto was more willing to make the journey himself. He expressed the hope that Newcomb would get to know Karl --who had studied some astronomy before beginning his diplomatic career -- and the new ambassador did, indeed, become a frequent and welcome visitor to the Newcomb household.

By June 1882, Otto had good reports on the progress of the objective, and finally committed himself to making the trip to America. He promptly began to worry about when to make it. He would need good weather for the tests, but he would not be free to leave Pulkovo until late in the autumn. What sort of weather could be expect, he asked Newcomb, in November and December? Should be perhaps postpone the journey until 1883? Then he returned to his old idea of combining this journey with an opportunity to observe the transit of Venus, until he realized that this would be just the worst time to see and to talk to his American friends and colleagues. Karl was planning a long holiday, and Otto wanted to be sure to be in Washington while his brother was in residence. In the end, the journey was made early in 1883; Newcomb was in Europe at the time (on his way back from observing the transit in South Africa). Hermann accompanied his father again; indeed, Hermann would be his "trusted deputy".[18] By then, Hermann had become deeply involved in theoretical optics. Some of his work on refraction is standard, and functions that he discovered during the course of this work are still known as "Struve functions". Newcomb had doubts whether a theoretical optician was the best man to test a telescope objective, to which Otto replied:

> You are completely right if you say, that theoretical knowledge of optics in itself alone is not enough for the testing, that for that [purpose] a skilled penetrating eye is still more important I agree completely with that and on that account have allowed Hermann, during the past year, to take part in refractor observations, especially however also in this year [I have] consulted him in the observation of more difficult objects under various atmospheric conditions, so that I believe he is also sufficiently prepared in practical respects [19]

Otto went on to say that, as far as he was concerned, he would accept the word of Dr. Draper, or of Newcomb himself, that the objective was acceptable. He was afraid, however, that if even the slightest flaw should later be found in the objective, he would be criticized for not having been present himself (or at least having sent a representative) when the objective was tested. All these doubts and problems were eventually resolved. Hermann and Otto left Pulkovo early in March 1883, made the tests, and returned on the *S.S. Bothnia* in April, communicating the results to Newcomb while still on board:

> In respect of our objective I can with full conviction take the view that it is essentially superior to the Washington [one] The colour correction is certainly more complete [word uncertain], the definition at least as good, if not better Of curvature we can discover no trace Whether the increase in light completely corresponds to the greater diameter could not be fully decided in a few evenings, since in this the temporary transparency of the air plays an important role, however the facility with which we saw the companion of the companion of Regulus, compared with the difficulty that the same offers, as a rule, in the Washington refractor, as well as details we noticed on some nebular surfaces speak decisively in favour of our objective Not quite acceptable to me is the "lap", as the Clarks call it, so very prominent in the middle of the crown-glass, but, to be sure, the light lost would be quite [un]important and diffraction fringes would become significantly obvious only for Sirius For the rest, the optical properties are so excellent, I think we certainly may console ourselves about the existence of this "lap" [20]

The temporary mounting that the Clarks set up for testing this objective did not display such good workmanship as was found in the lens itself. Otto gave Newcomb some advice on how to use it when he tested the Lick objective. He had suffered some accident while using the mounting and, even a year later, referred to "the false step that nearly cost me my life or at least a leg fracture..."[21] Newcomb brushes the matter off as a "slight accident".[22] A more immediate concern, however, was that Otto had failed to stipulate anything in his original contract of 1879 about the shipment of the objective to Pulkovo. Pickering agreed to accept responsibility for packing it and delivering it to a steamer bound from Boston to Hull. Once the objective was on board, it became Pulkovo's responsibility. In Hull, it would be transferred to another steamer bound for St Petersburg, under the supervision of a Swede, a former employee of the Clarks, to whom Otto offered a free return trip. Otto always found something to worry about and he confessed to Newcomb that "a load will be taken from my heart, however, when we first have evidence in Pulkowa that the objective has safely arrived."[23] The load was lifted on schedule; the objective arrived safely in Pulkovo in June 1883 --just as Otto had been led to expect.[24]

While he was in the United States, Otto was made an honorary associate of the National Academy of Sciences. The distinction clearly pleased him; he supposed that he owed it to Newcomb who was then Vice-President of the Academy. At the Naval Observatory itself, he had to be careful not to take sides in internal affairs. Traditionally, the Director was a serving naval officer, but a vacancy in that office created an opportunity for those who favoured having an astronomer in the position to make their views known. Otto's opinion was asked by several people, and there can be little doubt that he would have favoured the appointment of Newcomb, then in the Nautical Almanac Office, but he resolutely answered all those who approached him with a carefully worded formula:

> At the Naval Observatory, to be sure, a number of excellent works are carried out by individual astronomers; there is, however, a lack of coherence and system, and the latter will be attained in no other [way] than if a scientifically thoroughly educated astronomer will be appointed as Superintendent.[25]

Otto returned home again through England and Germany. In the latter country he visited Ludwig, who was then studying in Bonn under Schönfeld, Argelander's successor. Otto was godfather to Schönfeld's daughter, Anna, who recently published reminiscences of both him and Ludwig.[26] Otto also visited his sister Alexandra Kossman, in either Karlsruhe or Baden-Baden, and met Newcomb while in Hamburg for consultation with the Repsolds. Home at last, he hoped to prepare everything to bring the 30-inch into full operation by June of 1884. Again, delays occurred and again he began to worry. He was criticized for accepting the lens with the "lap" and had to explain that it would not affect the instrument's performance, without being able to demonstrate the fact. He found the delays in the construction of the dome frustrating and vented his feelings to Newcomb in April 1884:

> The [mounting] is already complete, upto the final balancing and, as it appears, Repsolds' are very pleased with the result. Probably J. Repsold will personally accompany this instrument here in June and supervise its installation, Now for that, we must also be ready with the dome; however, that we shall in fact be successful, I dare not say with certainty. From conversations with architects and engineers one can never get things right. In any case, even after the successful installation of the instrument, one or other will be on the dome and inside it to make improvements. Autumn could easily have arrived before that instrument is completely ready for work.[27]

The building for the 30-inch refractor (which no longer stands). Note the main part of the Observatory behind. (Reproduced from the same source as the photograph on p. 80).

The prophecy in that last sentence proved true, but a few months later Otto sounded more cheerful about the project:

> For two weeks now, our dome building goes forward briskly   The rails are already precisely levelled and centred   In 3 days shall also the other iron parts of the dome have been brought out and here all provisions are made so the installation of the same can proceed immediately   Correspondingly, I have written to Repsold that he may come here on Aug 15th to direct the installation of the instrument itself   Since, however, even after the successful installation, probably there will still be several things in the interior of the dome to be improved and arranged, I may not expect that the instrument will begin its proper activity before the beginning of October [28]

Otto gave up his customary holiday abroad in the summer of 1884 and, perhaps as a result, was suffering from gastric flu in October instead of testing his new instrument. He remained ill until January and vowed to go to Carlsbad or Vichy the next summer to take the waters --a cure to which he increasingly resorted as he got older. As it happened, he could have done little with the telescope during the winter of 1884-5. Pulkovo had unusually bad weather, and

Otto at the eyepiece of the 30-inch refractor --from a photograph of unknown date (probably c. 1885) in the private collection of Mr. P.G.E. Corvan.

the moving ladders and chairs needed for observing with so large a refractor were not sent from Repsolds' until the end of the year. At last, however, in February, Otto could write in a lyrical vein that echoed his father's description of Fraunhofer's Great Refractor in Dorpat (some sixty years earlier):

> The installation satisfies me to the highest degree; it combines ease of motion and convenience of use with complete solidity and an elegance of appearance, that does not leave anything to be desired. In fact it is a true masterpiece of technology and fully worthy of the objective, of the excellence of which we were persuaded two years ago in Cambridgeport.[29]

By the end of 1885, Otto could still sing the praises of his refractor to Newcomb.[30] It had survived the cold and thaws of its first winter and all parts functioned well. Plans were now afoot to electrify the telescope and dome machinery. Gill had advised this step, and Otto wrote to him too (in English) at about the same time:

> Our large refractor is a splendid instrument. With it the most difficult of Burnham's double stars are easy objects and Hermann has allready measured them all repeatedly. The mounting is a real masterpiece of the Repsolds. With all that the use of the instrument, on account of its colossal dimensions, requires great bodily exertions to which my age is not more adequate. For this reason its use will remain in the younger hands of Hermann.[31]

NOTES

1. M. Nyrén, 1905, Astronomische Nachrichten, Vol. 168, pp. 77-8.
2. O.W. Struve, 1889, Zum 50 Jährigen Bestehn der Nikolai Hauptsternwarte: Beschreibung des 30-Zölligen Refractors und des Astrophysikalischen Laboratoriums, St. Petersburg, Buchdruckerei der Kaiserlichen Akademie der Wissenschaften.
3. Simon Newcomb, 1903, Reminiscences of an Astronomer, (see Chap. 6, ref. 30), pp. 144-7).
4. O.W. Struve, 1879, letter of March 30 to Simon Newcomb.
5. O.W. Struve, 1879, letter of May 12 to Simon Newcomb.
6. O.W. Struve, 1879, letter of June 19 to Simon Newcomb.
7. O.W. Struve, 1879, letter of August 18 to Simon Newcomb.
8. Simon Newcomb, 1903, see ref. 3.
9. O.W. Struve, 1879, letter of October 27 to Cleveland Abbe. (Library of Congress, Cleveland Abbe Papers, Container 8).
10. O.W. Struve, 1879, letter of October 24 to Simon Newcomb.
11. O.W. Struve, 1879, letter of June 21 to S.W. Burnham preserved in the Mary Lea Shane Archives, Lick Observatory.
12. Most of the information about Anita Newcomb and her husband comes from The National Cyclopedia, 1909, Volume X, pp. 349-350.
13. O.W. Struve, 1879, see ref. 9.
14. O.W. Struve, 1879, see ref. 10.
15. O.W. Struve, 1881, letter of August 18 to D. Gill (see Chap. 13, ref. 29).
16. O.W. Struve, 1881, letter of September 1 to Simon Newcomb.
17. O.W. Struve, 1882, letter of January 13/1 to Simon Newcomb.
18. O.W. Struve, 1882, letter of June 28 to Simon Newcomb.
19. O.W. Struve, 1882, letter of August 28 to Simon Newcomb.

20   O W Struve, 1883, letter of April 30 to Simon Newcomb

21   O W Struve, 1884, letter of April 7 to Simon Newcomb

22   Simon Newcomb, 1903, see ref 3

23   O W Struve, 1883, see ref 20

24   O W Struve, 1883, letter of November 16 to D Gill (RGO 15/126 ff 824^r to 825^r)

25   O W Struve, 1883, see ref 20

26   Anna Tecklenburg, 1983, Die Sterne, Vol 59, pp 228-235 (edited by Th Schmidt-Kaler, p 233)

27   O W Struve, 1884, see ref 21

28   O W Struve, 1884, letter of June 19 to Simon Newcomb

29   O W Struve, 1885, letter of February 26 to Simon Newcomb

30   O W Struve, 1885, letter of December 29-30 to Simon Newcomb

31   O W Struve, 1885, Letter of December 29 to D Gill (RGO 15/126 f 851^{r-v})

CHAPTER 15

MAPPING THE SKY

By the time that the 30-inch refractor had been installed in Pulkovo and brought into full operation, Otto had turned 65. If he had thought of retirement before (Chapter 13), he must by then have been seriously considering it. He mentioned the matter again to Newcomb, in 1884:

> More than 10 months have passed since we met in Hamburg. Since then I have had a very busy time, in which I have not been much able to have quiet sessions at the work-desk or to make customary observations, being more claimed, indeed, by paperwork, journeys, conferences, etc. As the most important result of this activity I see, perhaps, the indications that I have succeeded in arranging the choice of Dr. Backlund as member of the Academy, in place of the deceased Sawitsch. We have by this not only gained a very worthy colleague, but I see in his appointment still more a guarantee that, when some day I shall no longer be in a position to direct Pulkowa, at all events, in the choice of my successor the scientific interest will be kept principally in sight. Of such an eventuality I must naturally think more each year.[1]

Indeed, Otto was thinking more of it during the next year and, writing of his illness during the winter of 1884-5 (Chapter 14), expressed himself even more openly to Gill. Complaining to the latter of feeling quite worn out, Otto continued:

> ...it might not be long before I shall feel compelled to follow old Airy's example and retire from office. A year or two might still elapse before that necessity arises, but hardly more. Perhaps I would do it even earlier if I was sure to get a good and trustworthy successor.[2]

Gill replied promptly, urging Otto to stay on: "you are still a long way from the age at which Airy retired"[3] he wrote. Perhaps, however, it would have been easier for Otto and Pulkovo if he had left in 1885. With the 30-inch telescope just completed, he could have retired full of honour and without enmity. Instead, when he did retire five years later, there was bitterness on both sides. In particular, by then Otto had quarrelled with his protégé Backlund, and no longer had a good word to say for the man.

Otto's illnesses became an increasing problem in his old age, and summer visits to German or Bohemian spas became almost necessities for him. He became an expert on European travel, and his advice was sought by Newcomb, who wanted *his* wife and daughters to stay in Europe while *he* was observing the transit of Venus in South Africa. Otto advised against their staying near the Lake of Geneva (except around Lausanne) or in Nice. "These places swarm with your countrymen and Englishmen... for the most part not of the best quality" he wrote.[4] Later he recommended Neufchâtel, at least as a centre for excursions:

> The language of the better society there is not, of course, strict Parisian French, but still
> that French which has been most widely spread throughout the world by the many tutors
> and governesses from western Switzerland. What the language of the middle and lower
> classes there is, is unknown to me.[5]

What the reaction of Grandfather Jacob would have been to that last remark is unknown to me. Despite Otto's help, Newcomb seemed always to have to pay more for accommodation than Otto expected. Perhaps, Otto said (in English in the middle of a German letter), they treated him "as an American with plenty of dollars in his pocket".[6]

In Pulkovo, Otto promised to give Hermann a substantial part of the observing time on the 30-inch (Chapter 14). The latter's second journey to America seems finally to have convinced him that he should become an astronomer --perhaps that was Otto's hope when he proposed that his son should accompany him. At any rate, Hermann took up the study of the satellites of Saturn and made it his own. Otto himself had observed the various satellites in the solar system --they offered a means, by a natural extension of the micrometer techniques that he had mastered in his double-star work, of determining the masses of the planets about which they revolve. Otto had observed the satellites of Uranus and Neptune (Chapter 9) as well as those of Saturn. The news of the discovery of the satellites of Mars, in 1877, by Newcomb's colleague Asaph Hall had, Otto wrote, been "a great treat"[7] for the meeting of the *Astronomische Gesellschaft* in Stockholm that summer. Thus, Otto probably directed his son into this field, but Newcomb, unwittingly it seems, gave an extra push, for Otto wrote to him:

> One of the first tasks that my son Hermann will carry out on this [30-inch] instrument,
> will be a series of observations of Hyperion [a satellite of Saturn], to which your newest
> brilliant work has still more prompted [him]. Hermann, who is quite inspired by your
> investigation has given us a most highly interesting long lecture on it. He is very sorry
> that you, as it seems from your letter, have viewed as a completed work the preliminary
> contribution in the Astr. Nachr. on the series of observations of Saturn's satellites begun
> by him. On this matter he will write to you himself.[8]

The study of a complex satellite system, such as that of Saturn, could lead not only to the determination of the mass of the planet, but also to the elucidation of interesting problems in celestial mechanics. If Hermann began by continuing a family tradition, he made this work very much his own, and later passed it on to his own son Georg. Our knowledge of the orbits of the satellites of Saturn and Uranus, especially, is firmly based on the work of these two Struves. They would, no doubt, have been astonished that space vehicles could be guided through these satellite systems and that of Jupiter --and perhaps even more surprised that their own work helped to make these feats possible.

Despite his talk of retirement, Otto kept busy well into the 1880s. In 1884, there was an international conference in Washington to consider which meridian of longitude should be adopted as the prime meridian from which all other longitudes should be reckoned. Otto did not cross the Atlantic a third time to take part in this conference, but he was keenly interested in its decisions. As everyone now knows, the principal one of these was the formal adoption of the, even then, nearly world-wide practice of using the Greenwich meridian. This won Otto's hearty approval:

> In respect of the resolutions of the Meridian Conference I take the standpoint that I consider it as a task incumbent on me to cooperate in their realisation with [the] best understanding. In particular, the resolution pleases me [that is] now ratified and probably soon coming into effect (even in France) to adopt the Greenwich meridian as prime, for the universal introduction of which I, in agreement with my father, have unceasingly striven for nearly 50 years.[8]

A second resolution, concerning the convention about when the day should be considered to begin, proved more controversial. Traditionally, astronomers have reckoned that the day began at noon. This was because the transit of the Sun across the meridian is observable, whereas the corresponding event at midnight is not (at least over most of the Earth's surface). For European astronomers, there was the added advantage that the date did not change at midnight and all observations made on a given night were made on the same "day". For most civil purposes the convention of beginning each day at midnight is obviously much more convenient, although navigators used the astronomical convention and international agreement was needed for any change. A resolution to make the change was adopted (some said pushed through, without adequate warning) during the Washington conference. Newcomb (not a delegate) opposed it vigorously, saying it would create confusion and even danger to navigation. He still opposed it in 1903, when he wrote his autobiography and records that Airy wrote to him:

> I hope you will succeed in having its adoption postponed until 1900, and when 1900 comes, I hope you will further succeed in having it again postponed until the year 2000.[9]

Gill, on the other hand welcomed the change:

> I was very sorry to see the persistent opposition of the German astronomers and Simon Newcomb to the change in the Astronomical day. If the Washington conference meant anything it was that we should all adopt Universal Time. Surely the Astronomer, whose daily business involves time, is better fitted to adopt the change of the beginning of the day than the comparatively ignorant public.[10]

Otto took a middle view, recognizing this resolution as a natural consequence of the other, but not wishing publicly to oppose his German colleagues. The change was, in his view, a sacrifice that astronomers should be prepared to make, but he agreed with Newcomb in hoping that it could be postponed until 1900. In December 1885, however, Otto believed that events were about to overtake them:

> What do you say, what does the Washington Observatory say, to the resolution of the Greenwich Board of Visitors, that for Greenwich and the Engl. Nautical Almanac, the change in reckoning Astronomical Dates shall begin on the 1st Jan. 1891? You know my

Kramskoi's portrait of Otto at the 15-inch telescope. (The original is in the Russian Museum, Leningrad.)

views well enough to know that in principle I am in agreement with the change, but the categorical action with respect to the epoch, if the same is undertaken, definitely has not my approval. Out of courtesy at least the English should have sought the views of other astronomers and directors of the calculation of ephemerides, even if they gave them no other weight than an expression of opinion to be discussed. Despite this I believe Pulkowa will follow the example of Greenwich, in order to avoid the greater confusion of not simultaneous changes.[11]

The resolution of the Board of Visitors was advisory only and was not acted upon, but --as Otto complained to Gill-- the apparent "want of courtesy in the proceedings [had] somewhat shocked him"[12], and he seemed inclined (unfairly) to blame Christie. It was a storm in a tea-cup: the change (still controversial) was not made until 1925, when Otto and Newcomb were both dead. Navigators pursued their business undisturbed.

At the end of 1885, Otto complained that he had become "an old man, quite tired of life"[13] although he still enjoyed many pleasures. One highlight had been a visit to Pulkovo by Alexander III and the Tsarina. A second had been Hermann's marriage to his second cousin, Olga, another great-grandchild of Jacob Struve and the daughter of the elder Hermann Struve (son of Wilhelm's brother Ludwig) who lived in Riga. The astronomer Hermann had also completed the first phase of his observations of Saturn's satellites and was working hard on their reduction. Otto expressed pleasure at his son's progress. Hermann had been offered two professorships, each at higher salaries than he received at Pulkovo, but Otto correctly predicted that Hermann would decline them both. Ludwig, too, was pleasing his father by his progress. In this same year of 1885, Newcomb had paid a second visit to Pulkovo, in company with Anita. During this visit, Otto had suggested that Newcomb should have his portrait painted (at Pulkovo's expense) and hung in the rotunda of the Observatory. By October 1886, the portrait had arrived and Otto wrote that it

...is quite excellently well done, only you have in it a more earnest expression than is normally habitual for you. It will have its place in our hall by the side of that of Hansen.[14]

A portrait of Otto had also been made during 1885. It is one of the more important portraits of him, which perhaps lends interest to his rather non-committal comments on it:

Also my portrait has been executed in oils during the past year, (not for the Observatory but for the family on Döllen's order); it has, however, found some very diverse judgments. Worked on with special pleasure by Kramskoi, the best portrait-painter of Russia, it is in artistic respects certainly quite beyond reproach. But the unaccustomed attitude and lighting (the artist had painted me sitting before the telescope and with a dazzling light from a lamp at night) had made the judgment[s] on the likeness, and particularly on the expression of the face, turn out very variously.[15]

From portraits of astronomers, we turn naturally to portraits of the sky. Just at this time, it was becoming possible to apply photography to astronomy, and leaders in this field were the brothers Paul and Prosper Henry in France and David Gill in South Africa. Gill had photographed a bright comet in 1882 (known as the "Great September Comet") and was impressed by the number of stars recorded on the plates. He had the idea of making a photographic survey of the southern sky to complement Argelander's northern *Durchmusterung*. He set to it with his customary energy and, in early 1886, wrote:

> Our star photographs are going on quite rapidly, and I have hopes soon of being able to tell you that the measurement and reduction of the plates has begun, and to give you some idea as to when the Southern Durchmusterung may be expected to be completed.[16]

Meanwhile the brothers Henry, had won the support of Admiral Mouchez, then the Director of the Paris Observatory, and were preparing an international conference to discuss the possibility of all major observatories joining in the making of a photographic *Durchmusterung* of the whole sky to a much fainter magnitude, of course, than Argelander had been able to reach. Otto was interested but cautious; his letter to Newcomb in February 1886 hardly foreshadows the extent to which he would become involved:

> In these days I have been very much occupied by the Paris star photographs and am full of admiration for these [and] what the Messrs. Henry are in a position to do. In this way a great reform in many parts of practical astronomy could easily be brought about. The powerful refractors will to some extent lose their significance, if, as it appears, the photographic plate is so much more sensitive than the human eye. But however [far] this sensitivity reaches separately, it will not replace the telescope, at least not at present. Of this we have a proof in the nebula surrounding the star Maia in the Pleiades. Certainly the existence of this is recognized on a photograph exposed with a 12-inch refractor, but the latter hardly gives an idea of the form of the nebula, while our 30-inch quite distinctly displays various details. With the 15-inch hardly a trace of the nebula is recognized. Mouchez has the thought now to prepare [by] the combination of the power of many observatories, by photographic means, a complete chart of the sky (which should contain all stars to at least 15th magnitude) and I admit that this idea pleases me very well. First, however, for the participating observatories similar instruments must be made and strictly-to-be-followed methods, standards, etc. must be laid down. The first has its difficulties on account of the not insignificant costs, the other requires in any case still many studies. Probably, for this purpose, Mouchez will invite the "leading astronomers" [English phrase in a German letter] of the various countries to a general council in Paris, and I would even not be unwilling to accept such an invitation. You or even Holden should also come to it. Also for Pickering the matter may be of interest, since obviously photography promises to play a role as a photometric device.[17]

Gill strongly supported the proposed conference (or perhaps even had the same idea independently) and suggested to Mouchez a date in 1887, when he (Gill) planned to come to Europe anyway. In the summer of 1886, as usual, Otto travelled west. He had to represent the St Petersburg Academy at the 500th anniversary celebrations of the University of Heidelberg. Bischoffsheim also wanted Otto's opinion of the now completed refractor at Nice. Thus Otto had a good opportunity to visit Mouchez and to discuss with him the proposed conference. As to the Nice refractor, Otto had to some extent made up his mind before he went:

> In respect of the mounting I do not doubt a priori that our instrument significantly
> surpasses the Bischoffsheim [one].   In fact, I cannot boast enough, how carefully the
> Repsolds devised and executed all the parts of our refractor.[18]

(Or as Gill put it, "Their obvious desire is to make the best possible instrument
and to be more careful of their reputation than their pockets."[19]) Otto was
impressed by the Nice refractor, but he did not significantly change his opinion
after his visit. In Paris, he and Mouchez agreed on mid-April of 1887 for the
conference on the project that came to be known as the *Carte du Ciel*. Otto
feared that too many people would come for the conference to be easily
manageable, but he thought their mutual stimulation would be useful even if
they reached no conclusions about the principal topic. What would he have
made of a modern I.A.U. General Assembly with 2,000 participants?
Characteristically, he was beginning to hesitate about his own participation in
the Paris conference and proposed sending Hermann and Hasselberg. Gill tried
to persuade Otto to come himself, not --as it seems-- solely for scientific
reasons:

> My wife and I both rejoice that you are still so well and able for long journeys, and no
> doubt for many a cigar as of old. I can join you now in smoking many more than formerly,
> and am longing to look at you again through a veil of thin blue smoke.[20]

The decision of whether or not to go was taken out of Otto's hands, as
he afterwards explained to Newcomb:

> Against my will I was compelled by the Academy (which perhaps was guided by ulterior
> motives) to appear personally at the Paris astrophotographic congress.   That I as
> President of this Assembly contributed something to dampen the rashness of the optimism
> of those who see photography in its present state already as the non plus ultra of practical
> astronomy and, on the other hand, strove to give this undoubtedly powerful resource the
> necessary further development and rigorous foundation, of this the Procès-verbaux of this
> assembly will give you some proof. Naturally my task as President was no easy [one] and
> it was three hard-working weeks that I spent in Paris, and likewise naturally, many
> honours (whether deserved or not, we will not investigate) were given to me there. The
> last have had the unpleasant consequence that they called forth jealousy and ill-will
> amongst my academic colleagues, which expressed itself in that, immediately after my
> return they passed a resolution, that compelled me to the declaration that under such
> conditions I can no longer remain Director of Pulkowa. The conditions must therefore be
> changed, if I, as at least by all parties it was wished, shall remain in my position and for
> that a change in our regulations is necessary. Such a change, however, can come about
> only with the cooperation of the President of the Academy, Count Tolstoy, [not the writer
> Count Leo, but the statesman Count Dmitri, at that time an influential minister and
> Otto's friend] and since May he has been sick on his estates in the interior of the Empire.
> Thus the whole of Pulkovo, with me, was put in a most awkward state of uncertainty, that
> crippled all our work and business. At the same time the concerns of the Solar eclipse, to
> which I fully gave myself, were a real diversion for me; therefore now in this question a
> decision must come.[21]

Otto was in a mood to see change and decay all around, as the next
paragraph of the same letter shows:

> In Germany also unfavourable times for astronomy appear to have broken out. Recently the very insignificant Becker from Gotha has been appointed Director of the Strasbourg Observatory [to succeed Winnecke] and to his place in Gotha will go, as it is rumoured, a still more insignificant man. In both these appointments, by these [rumours], as it appears, women had the deciding voice.[21]

Otto's retirement is the subject of the next chapter: here we must return to the *Carte du Ciel*. Otto's appointment as President of the conference was an appropriate recognition of both his personal seniority and of the importance of the observatory that he directed. He had already, by his publications, demonstrated his interest in the subject, and he was made a member of the *Comité permanent* that was to supervise the execution of the Conference resolutions. Of these, the principal was to construct a photographic chart --by international cooperation-- of all stars down to the fourteenth magnitude. Since, on these plates, the images of many bright stars would be too large for accurate measurement of their positions, it was also proposed to obtain a second series of plates showing stars only to the eleventh magnitude.

David Gill, after his enthusiastic participation in the Paris conference, went on to England where he encountered much opposition. This is made clear by several articles in *The Observatory*, at that time edited by H.H.Turner and A.A.Common. Gill had understood that the Paris conference had resolved on the creation of a *catalogue* of all stars down to the eleventh magnitude. The Editors of *The Observatory* thought this impractical and unnecessary and, even though Admiral Mouchez himself wrote in support of Gill's interpretation, they maintained that Gill had misunderstood the second resolution of the Conference, that other participants would not have supported it if they had shared Gill's point of view, and that a catalogue was not needed since any astronomer would be able to refer to the original plates.[22] It is difficult to avoid the impression that the Editors were at least partly motivated by personal considerations, especially since Turner was strongly committed to the *Carte du Ciel* himself. Worse still for Gill, he discovered that the commmittee of the Royal Society on which he relied for support for his own southern *Durchmusterung* considered there was no need for this as well as the *Carte du Ciel*.[23] He appealed to Otto for help, and received the following reply:

> First of all I must plead guilty with regard of my not having written you before this. In fact I did not expect your affair would some so soon before the R.[oyal] S.[ociety] Committee. Besides I did not think it favourable for the contents of the letter, to write it in the hurry of a rapid journey. It was only last week I came to Berlin, that [I] became acquainted by Auwers with the state of things, and that was rather too late for me to write.

> You were certainly quite right to express to members of the R.S. Committee, and particularly to Stokes, that your photographic work, the Durchmusterung, had my full sympathies and is regarded by me at the present time the most useful and important undertaking for astronomy of the Southern Hemisphere.

> Therefore I am very sorry your application to the Government (Instrument) Grant Committee has met with such unqualified opposition. Christie's proceedings are a disgrace for English men of science. If he continues to proceed, as he has done in this matter as

well as in the time question, the authoritative position of Greenwich in the astronomical world will be [soon] lost   No foreign astronomers will have to do with [envious opinions] quoted by a man who abuses of the position he has accidentally won (not by his merit), to destroy general harmony between astronomers, and suppress useful work undertaken by men with whom he cannot compare in scientific merit and understanding   Poor old Airy, the position he has won for Greenwich by more than 50 years [actually, slightly less] earnest working and his highstanding character will be lost almost irreparably   Sic transit gloria mundi!

With these feelings concerning Christie, I am then very glad you will continue your photographic work at the Cape with all the means [at] your disposal   You may be sure that your undertaking [is not] only approved by all foreign astronomers but will be assisted by them wh[en]ever convenient occasion will offer itself [24]

Otto wrote this letter in the midst of his own personal crisis in Pulkovo. and this may explain why he was so scathing about Christie.   No public evidence links the latter with the attacks on Gill in *The Observatory* and Christie had at first supported Gill's efforts at the Cape.   From this time on, however, Christie was hostile to Gill.[25]   Eventually, the Royal Society Committee under Sir George Stokes --the famous physicist and then President of the Society-- did approve a photographic telescope for the *Carte du Ciel* work at the Cape, and Gill was able to continue with his *Durchmusterung* (which turned out to be the more successful of the two projects) by financing it himself.   With hindsight, we can see that both sides in the dispute in the pages of *The Observatory* were partly right.   A century later, we still use star catalogues, but the *Carte* was an ambitious project, undertaken before its time and never completed in the way originally intended (although the Cape contribution was amongst the best).[26]   The task that it was intended to perform has since been done better by the Palomar Schmidt telescope in the northern sky and its twin in Australia in the southern.   Otto himself had reservations about the project, and these grew during his retirement, when he remained a member of the *Comité permanent*.   On the other hand, the project was a major effort in international cooperation, and like the observations of the transits of Venus, helped to teach astronomers how to organize themselves for such work. Nearly two decades after Otto's death, the International Astronomical Union was founded and one of its standing commissions was concerned with the *Carte du Ciel*.   The transit observations, the 1887 conference and the *Astronomische Gesellschaft* were all precursors of the Union, and Otto was intimately involved in all three.   He surely would have rejoiced to know that his grandson and namesake would one day become President of the International Astronomical Union.

NOTES

1   O W Struve, 1884, letter of April 7 to Simon Newcomb

2   O W Struve, 1885, letter of December 29 to D Gill (see Chap 14, ref 31)

3   D Gill, 1886, letter of January 28 to O W Struve (RGO 15/126 ff 852^r to 858^r)

4   O W Struve, 1882, letter of June 28 to Simon Newcomb

5   O W Struve, 1882, letter of August 28 to Simon Newcomb

6   O W Struve, 1883, letter of February 6 to Simon Newcomb

7.   O.W. Struve, 1877, letter of December 7 to Simon Newcomb.

8.   O.W. Struve, 1885, letter of February 26 to Simon Newcomb.

9.   Simon Newcomb, 1903, <u>References of an Astronomer</u>, (see Chap. 6, ref. 30), pp. 227.

10. D. Gill, 1886, see ref. 3.

11. O.W. Struve, 1885, letter of December 29-30 to Simon Newcomb.

12. O.W. Struve, 1885, see ref. 2.

13. O.W. Struve, 1885, see ref. 11.

14. O.W. Struve, 1886, letter of October 10 to Simon Newcomb.

15. O.W. Struve, 1886, letter of July 5 to Simon Newcomb.

16. D. Gill, 1886, see ref. 3.

17. O.W. Struve, 1886, letter of February 10 to Simon Newcomb.

18. O.W. Struve, 1886, see ref. 15.

19. D. Gill, 1886, see ref. 3.

20. D. Gill, 1886, letter of September 14 to O.W. Struve (see Chap. 13, ref. 28).

21. O.W. Struve, 1887, letter of August 30 to Simon Newcomb.

22. D. Gill, E. Mouchez and the Editors, 1888, <u>The Observatory</u>, Vol. 11, pp. 205, 224, 255, 296 and 320.

23. D. Gill, 1896, <u>Cape Photographic Durchmusterung, Annals of the Cape Observatory</u>, Vol. III, p. XV (Introduction).

24. O.W. Struve, 1887, letter of May 23 to D. Gill (RGO 15/126 ff. 268^r to 269^r). Gill himself copied out this letter and it is his copy that survives. Some words are made illegible by blots on the original.

25. B. Warner, 1979, <u>Astronomers of the Royal Observatory, Cape</u>, A.A.Balkema, Cape Town and Rotterdam, p. 93.

26. P. Couderc, 1970, <u>Transactions of the International Astronomical Union</u>, XIVB (C. de Jaeger and A. Jappel eds.), D. Reidel, Dordrecht, pp. 172-8.

CHAPTER 16

OTTO'S RETIREMENT AND LAST YEARS

Earlier chapters have provided evidence that Otto had been toying with the idea of retirement for some years before the crisis that faced him on his return from the *Carte du Ciel* conference; he just seemed unable to make a decision. He celebrated his sixty-eighth birthday a few days after his return home; only eighteen months earlier he had had a long illness during which, as he admitted privately to his friends, the administration of the Observatory suffered. It was clear that whatever happened, Otto must give up the directorship soon. Many members of the Academy may have thought it best, both for Otto and Pulkovo, to precipitate the decision. To this extent, Otto may have been right in accusing them of "ulterior motives" (Chapter 15), although he may have misread those motives for something worse than they were. Nevertheless, the situation was a complex one with many factors affecting the course of events.

Foremost among these factors was the well-known division in Russian society of that time between *westernizers* and *slavophiles*. Peter the Great had first opened the country to westerners (often called "Germans" in Russia, even when they came from other countries) and Catherine the Great continued his policy. Although the Struves did not come to Russia until the time of Alexander I, they were typical of families who came and kept their language and culture while beginning to think of themselves as Russians. The old Empire, like the modern Soviet Union, contained many nationalities and language groups and it no doubt seemed natural to immigrants to form new such groups. Otto did speak Russian, but not well, and although born in the Empire had had to apply for citizenship. He nevertheless made his own feelings of nationality clear; for example, when Newcomb told him how popular Karl had become in Washington, Otto wrote that the news made him "as brother and as a good Russian patriot" very happy.[1] To a considerable part of Russian society, however, Otto was a foreigner whom they wished to remove from power. He can hardly have helped his own position by having appointed so many Germans and Scandinavians to vacancies in the scientific staff. By themselves, the slavophiles might not have removed Otto, who had to leave soon anyway, but they certainly played a role in the choice of his successor.

A second factor was that Otto had been in charge long enough. In 1887 not only was he, even in his own eyes, an old man, he had been Director for 25 years and in charge, *de facto*, even longer. Such was the continuity between father and son, some might almost have thought that Pulkovo had had only one Director since its foundation. If, at the beginning of Otto's term, much was "said of Poulkova as a family affair" (Chapter 12), towards the end he regarded Hermann as his natural successor. The latter received a large share of

219

observing time on the new 30-inch, and although he undoubtedly used it well, such generous allocations to a relative newcomer were almost bound to cause resentment, especially amongst those who thought the large instrument should be used for modern astrophysical research rather than for something that, even then, was beginning to seem old-fashioned.

A third factor was that the subordination of Pulkovo to the Academy, although a natural arrangement, had often been a source of friction. Wilhelm had found the Academy difficult to work with after the death of Paul Fuss (Chapter 11) and Otto had problems with it all through his tenure of office. It is against this background that the decision taken by the Academy while Otto was in Paris seemed so objectionable. To neither Gill nor Newcomb does Otto make quite clear what this decision was. Nyrén seems to suggest that it concerned a new staff appointment --the Academy's formal approval was needed-- but he is not explicit.[2] Nyrén does say:

> The request in question was not in itself important enough to give grounds for so important a decision [i.e. Otto's resignation]; those on the opposing sides were doubtless well aware of this. The deep grievance which Struve felt in the refusing vote of the Academy came from his conviction that they wished to strike himself.[2]

If Otto had retired voluntarily when he had completed the 30-inch refractor he might, despite the fact that the pressures just described were already beginning to be felt, have left with all honour and without bitterness. He might even have had a say in the choice of his successor. Unfortunately, in 1887 Otto came to feel that he was being unceremoniously pushed out after a lifetime of service and freely expressed his bitterness in his private correspondence: a bitterness to which Hermann explicitly referred in a letter he wrote to Newcomb shortly after his father had retired.[3] This bitterness was focussed on Backlund, the erstwhile *protégé*, whose election to the Academy Otto had himself been at some pains to arrange (Chapter 15). Shortly after his election, Backlund exercised his right as an Academician to leave Pulkovo and to live in St Petersburg. As late as February 1886, Otto wrote to Newcomb -- apparently with approval-- about Backlund's work on Encke's comet. About this time, however, Backlund completed work on Volume 8 of the Pulkovo *Publications* -- one of the star catalogues that was the result of some of the most fundamental work that Wilhelm had planned. Otto was dissatisfied with Backlund's introduction to this volume and substituted one of his own. Backlund then began to publish criticisms of Pulkovo's observations and reductions, and Otto published rejoinders. Whether or not the personal animosity was mutual is unclear, but in many of his letters Otto became quite scathing about Backlund, and appears to have believed that the latter was scheming to become his successor and was supported and encouraged by Gyldén --which last seems unlikely, since Gyldén (who had taught Backlund) had returned to Sweden in 1871.

News of Otto's intention to resign in 1887 eventually reached the Tsar (Alexander III) who interviewed Otto personally and asked him to remain in office until the Observatory celebrated its 50th anniversary in August 1889. After that celebration, Otto was to say whether or not he still wished to leave.[4] This decision of the Tsar's was the more remarkable in that Alexander was

Photograph of Oskar Backlund (reproduced by permission of the Royal Swedish Academy of Sciences).

much influenced by the slavophiles and had the reputation of disliking "Germans". Traditions and anniversaries were important to Otto and still to be in charge of Pulkovo for its Golden Jubilee must have meant a lot to him. He seems, however, to have decided that he should go immediately afterwards, for, as he wrote to Newcomb:

> ..I feel only too much that my strength is no more growing to the measure of the tasks of the position, as I must wish for my own satisfaction. What is more, through the grace of the Tsar, I will probably be placed in the position to work decisively in the appointment of my successor, whereas before the antagonism of the Academy not only was trying to cut me off, but even had foisted [on Pulkovo] a successor who was not even qualified for the position.[4]

In the circumstances, as Otto wrote to both Newcomb and Gill, the jubilee celebration was bound to be muted --if for no other reason than that it was arranged on short notice. Foreign astronomers were invited. Neither Newcomb nor Gill could attend and Abbe did not receive an invitation (although Otto did write to him explaining why he had not been sent one). A book was compiled for the occasion, which contained Otto's official account of the construction of the 30-inch refractor.[5] Foreign observatories, even if they could not send representatives, were asked to send congratulatory addresses that might be of help to Pulkovo in the future. David Gill obliged with both a formal address and this personal letter:

My Dear Struve

Herewith, by registered parcels post, we send you a greeting from the Cape in shape of our formal congratulations on the celebration of your 50th anniversary in Pulkowa.

What we have said formally is the barest and driest of simple unvarnished fact. We have no higher praise to give, there is none that can be given higher than the simple statement of what Pulkowa has done for Astronomy. But for my wife and myself there is in our hearts a warm and deep feeling of sincere friendship gratitude and esteem -and also a feeling of sympathy in the troubles which recently have unfairly surrounded you.

But when a man can, on an occasion like this anniversary, point to a record such as yours and think of the Observatory which you have directed --he can afford to forget the petty worries which he may have met with in his career and say proudly "these things I have done and this is the Institution which I have directed.["] Few men could stand more proudly than yourself to face a just tribunal on such grounds. If you should retire from Pulkowa it will certainly be to the regret of all those who love Astronomy for its own sake. But you will be able to lay aside your work with a satisfactory feeling of having nobly done your duty, and of having contributed a noble share directly and indirectly to the advancement of Astronomy.

Remember me most kindly to all my friends on yr [i.e. your] staff who may remember me --specially to Döllen, Nyrén and Romberg-- also remember me to such of our colleagues who more fortunate than myself are with you at yr celebration.

My wife desires to join me in all kind messages. Auwers sends his own.

         Believe me

            always dear friend

             Sincerely yours

               David Gill.[6]

Otto showed his appreciation of this letter by replying in kind, after the celebration, and giving Gill a brief account of what had happened at it:

My Dear Sir,

Accept my warmest thanks for your kind letter and the excellent address, by which you and your staff have congratulated us at our festival. I dare say the address has made the most beneficial effect upon all, who have read it.

Our festival is over and had pretty well succeeded. About 300 people have come from Petersburgh to attend the meeting and I expect it will not remain without favourable results for the Observatory. You will easily understand that on account of the circumstances (the strange relations to the Academy) we did not send out many invitations. Then more we were rejoiced by the fact, that some of our friends did not refrain from the long journey to join us on the festival. Schönfeld, Repsold, Christie, Hasselberg and Pechüle were come, not to speak of the Russian astronomers. It will interest you to hear that we have found Christie here quite different of what he formerly used to be. During a weeks stay he appeared quite a modest man, without any pretensions except that to learn and become acquainted with what other people have done. May be, the heavy loss, he has experienced last spring, by the death of his wife, has acted upon his temper.

Next month the struggle with the Academy (more exactly with the perpetual Secretary Mr. Wesselofsky and his helpmates Wild and Backlund) will begin anew, but this time, thanks to the friendly intentions of the Grand Duke President, with much more favourable prospects. I expect that during the winter I will obtain for the Observatory that freedom of action, which in my opinion is necessary for its success. This [once] obtained I shall think myself entitled to retire from business. I feel myself tired and long for repose. It will be than [sic] time to entrust the direction of the Observatory to younger hands.

On account of the imminent struggle I cannot leave the Observatory during the next months and therefore cannot attend the meeting of the (I beg leave to say "foolish") Comité permanent de la Carte du ciel. On my place I have sent there Mr. Nyrén. What will be the result of the meeting I have not the least idea. To defend the interests of exact astronomy, there will be only Vogel, Bakhuyzen and Nyrén and they will have to contend with two or three dozen of Frenchmen and their American followers. Also Christie does not look very hopefully upon this meeting.-- Mouchez has written to me that immediately after the meeting he will retire from service. Who will be his successor, Tisserand or Leowy? I am uncertain which of the two I should prefer in that position.

* * * * *

Thanks to your dear lady for her kind remembrance. I hope she is now quite well. Please convey to her Döllens and mine heartiest compliments.

Believe me, dear Gill

Every truly yours

Otto Struve.[7]

Otto had already been to Paris once that spring, to make a speech at the unveiling of the statue of Le Verrier in front of the Observatory. Count Tolstoy had died and was succeeded as President of the Academy by the Grand Duke Konstantin . If this had given Otto more reason to hope that he might have some voice in the choice of his own successor, he was to be disappointed. Newcomb suggested Hermann as the obvious person and, thanking him, Otto confessed that he agreed but added "...I cannot take the initiative in it, just because he is my son and it is in the interest of *certain* Adacemicians that Pulkowa should not have the most suitable Director".[8]

There were two counts against Hermann; he was a Struve and a "German", and he did not get the job. Accounts of his own life suggest that he did not want it just then --he was still too busy with Saturn's satellites to take on the burden of administration, although there may have been some understanding that he would eventually succeed.[9] The new Director, appointed in the spring of 1890, was the indisputably Russian astronomer Bredichin, from Moscow --in which city he was allowed to continue to live while a Vice-Director (expected to be Nyrén) looked after the Observatory's day-to-day work. Otto, writing to Newcomb from Switzerland, made clear that he thought little of the arrangement, but he hoped that Nyrén, Hermann, Wittram and Renz could hold the Observatory together --now he himself was removed as an object of animosity-- until better times. "At least" he growled "the scoundrel Backlund has not obtained his hoped-for reward".[10] In this same letter, he poured forth all the bitterness and frustration he felt in the manner of his departing. Although he tried to maintain that he had left because he no longer had the strength to do the job to his own satisfaction --and this may have been true as far as it went-- everything else in the letter indicates his profound personal unhappiness.

The family broke up, too, when Otto left Pulkovo. All the children of his first marriage had grown up, the sons launched in their own careers and the daughters married. Döllen did not wish to work in Pulkovo under another director but, not being ready to retire himself, found a university post in Dorpat. Ludwig, too, after his period with Schönfeld in Bonn, Schiaparelli in Milan and further study in Leipzig, also came to Dorpat as an assistant in the observatory that his grandfather had made famous. There, just about at this time, he undertook his principal astronomical research: a repetition of his father's investigation of precession and solar motion, with more modern data. Eva, the sole child of the second marriage and by then in her twenties, had been sent a few years before to a finishing school in Lausanne. Thus, Otto and Emma had some difficulty choosing where they would live, although it seemed obvious enough to spend the first winter (1890-91) in Switzerland and Rome. The following October, however, saw them back in Russia, in Tsarskoe Selo -- Otto could not face a visit to Pulkovo itself. Two reasons brought them back: the sale of his personal library (Chapter 12) and his wish to complete Volume X of the *Pulkovo Observations* --his own measurements of double stars. He was pleased to learn that Hermann, encouraged by Newcomb's praise of his work on Saturn's satellites, was still left undisturbed on the 30-inch refractor, but less pleased that --instead of Nyrén-- a Russian, Sondoff, "in astronomical matters a completely unknown quantity"[11] was Vice-Director and other affairs at the Observatory also disturbed him:

> As a specially pregnant fact I will mention that in the spring Bredichin has dismissed from the Observatory the two most skilful supernumerary astronomers that had come in my time, Blumbach and Wanach, without any preparation, by giving them as reason the frank explanation "He cannot use them since they do not have Russian names". They are both poor devils; at least for Blumbach there is Vogel's recommendation in Potsdam, where he worked well for a year... The other, Wanach, does not see where he will get a piece of bread.[11]

In fact, Wanach also found a post in Germany, helped by the Becker in Strasbourg whom Otto had dismissed so scathingly (Chapter 15)! Otto stayed in Russia well into the summer of 1892, partly because Hermann's only son Georg (one day to become an astronomer) had been severely ill with diphtheria during the preceeding winter. Otto visited his married sister in Karlsruhe later that summer, however, and also saw Newcomb who was in Europe again.

Sometime in the autumn or winter of 1892-3, Otto went back to Russia primarily to continue his work on Volume X of the *Pulkovo Observations*. He wanted to have this volume published on the hundredth anniversary of his father's birth, in April 1893, but the printer delayed the publication. Döllen planned a special publication for that day and Ludwig hoped, on the same day, to release his observations, made in the previous year, of occultations of stars by the totally eclipsed Moon. The day itself (April15/3) was marked in Dorpat by Arthur von Oettingen's commemorative lecture, extensively quoted in earlier chapters. Otto, his sons, and Döllen were all present. Some time ahead of the anniversary, Otto wrote of it to Newcomb:

> For me it is a fond wish to spend that day in Dorpat in company with Döllen and my son Ludwig and even in the same rooms in which I grew up and in which Father created his great works Since New Year, of course, Ludwig as Observer has lived in the residence of the Observatory Director, because Schwarz, who without resigning his position and without being able to satisfy [the duties of] the same, from imaginary considerations of health has given up this official residence and has built a private one Thereby Ludwig's prospects of becoming Schwarz' successor have still not essentially improved The mood against the German element in the Baltic provinces is still unchanged as the temporary dictator Indeed, it has recently baptised my native "Dorpat" as the Russian "Jurjevo" That, however, does not hinder any almost immediate answer about the 15th April safely arriving for me at the address "Dorpat Observatory" since abroad, on the contrary, this change has not yet taken hold and there, indeed, the name Dorpat will long keep its good ring [12]

Ludwig, in fact, despairing of advancement in Dorpat was actively looking for positions outside Russia. His father wrote to Gill on his behalf, who replied that he wished he could help the grandson of Wilhelm ("who must have been such a glorious fellow in every way"[13]) but could not since all permanent positions at the Cape were filled by the Civil Service Commissioners in England. America, Gill thought, offered Ludwig the best prospects. In fact, Ludwig did not leave the Russian empire but became, two or three years later, the Professor of Astronomy and Director of the Observatory in the University of Kharkov.

Later in 1893, Otto and Emma finally decided to live, with Eva, in Karlsruhe, although Otto made fairly frequent summer trips to Russia until the end of his life. Early in 1895, a new crisis occurred in the affairs of Pulkovo , which Otto described in full to his friend Newcomb·

> Bredichin, who for five years has undertaken the directorship of Pulkovo with reluctance, since as a rich man he did not need the position and shrank from its problems, announced at Christmas that he would resign the office and return to his estates As his successor he proposed primo loco and ex aequo, Backlund and my son Hermann, but the administration

that already for a long time [has made] the Academy, on whom unfortunately the filling of the position depends, hostile to the Struves, determined [at once] that for the present only the first would be subjected to ballot. By that, of course, the matter was, in a manner of speaking, decided and Hermann at once pronounced he was resolved to give up his position at Pulkowa, since it was against his honour to remain in it under Backlund's directorship.[14]

So, after all, Otto had to swallow the (for him) bitter pill of seeing Backlund in charge. While the formalities were being completed, Hermann was offered the directorship of Königsberg Observatory through the mediation of Auwers. Königsberg, now the Russian city of Kaliningrad, was then still Prussian, and it was Hermann, not Ludwig, who left his native country. Thus, appropriately, Wilhelm's grandson came to occupy Bessel's chair. Hermann proved to be probably the best director Königsberg had had since Bessel's death. and his success there soon led to his being offered the more important post of Director of the Berlin Observatory. Meanwhile, when his Russian colleagues realized he was serious about leaving the country, they tried to persuade him to stay, even offering him the directorship of a new observatory (still to be built) in the Crimea:

Such temptations made him doubtful for a couple of weeks, whether he was doing right to leave Russia, but he soon convinced himself that it is easier to make promises than to keep them and, by that time, he had some days before definitively accepted the call to Königsberg. Probably he will take up the new office already on May 1st. I hope and believe that he has the strength in him to raise up again the observatory [that] has declined so far since Bessel's death.[14]

Apart from these concerns, Otto clearly enjoyed his early years of retirement. In the climate of southern Germany he could take exercise in the fresh air throughout the year. Although in his late seventies, he kept active and read both the *Astronomische Nachrichten* and the *Monthly Notices*. As late as 1898, for example, his interest piqued by a paper in the former publication whose author, Cerulli, had a theory about the Martian "canals", Otto wrote to Schiaparelli "the highest authority in such things, to *you* my dearest friend, with the request to send me your opinion on this announcement".[15] Schiaparelli, having coined the term "canali" which had been mistranslated into English as "canals" (thus causing confusion and controversy and, incidentally, inspiring numerous works of science fiction) had indeed some claim to be called "the highest authority in these things". He was not impressed by Cerulli's claim that the surface of the Moon showed linear features resembling the canals and replied (in French) "I read your lines with the greatest pleasure, because they are the most convincing proof of your state of good health... You do too much honour to the idea published by my estimable friend Mr. Cerulli, even Homer nods [*...quandoque bonus dormitat Homerus.*]"[16] Yet, in an odd way, Cerulli has been more nearly vindicated than Schiaparelli would have supposed possible. The Martian "canals" are not genuine features but there are resemblances between the surfaces of Mars and the Moon.

Otto also discussed astronomy in his letters to Newcomb, particularly the latter's own work on the fundamental constants, the founding of the American Astronomical Society and, once again, the progress of the Lick Observatory.

Otto's first principal task after settling in Karlsruhe, however, was writing the biographical account of his father -the *Erinnerung*-- used extensively as a source in earlier chapters. It was intended primarily for family and close friends, and not as a formal biography. Nevertheless, despite some errors of dates and detail, it remains a valuable source. Newcomb was amongst the close friends who received a copy, as Otto's letter of 12 November, 1895, makes clear.

Newcomb and Gill both planned to come to Europe in 1896 for a meeting in Paris concerning the *Carte du Ciel*. Otto, although still a member of the *Comité permanent*, did not intend to go. He was alienated from the project and felt that "the excitement of a stay in Paris at such a time would hardly do me good".[17] Besides family affairs took him to Russia and Königsberg. Thus he missed a chance to see his two old friends again, but he wrote to Newcomb afterwards:

> We have, in fact, spent a very pleasant summer marked both by happy events and uninterrupted fine weather. At the beginning of June we celebrated in Dorpat the green wedding of my son Ludwig, then at the beginning of August in the dacha of my son Wilhelm, in the surroundings of Petersburg, my own second silver wedding and finally, after the return from Russia, I made still one detour to Chur, in order to be present at the golden wedding of my friend Herold.[18]

Pastor Herold appears to have been a very old friend and a descendant of Johanna Bartels' ancestors. Otto introduced him to Newcomb, on one occasion, but the name never appears in Otto's letters to the latter until about this time.

The week in Königsberg had been especially enjoyable for Otto, because it broke the scientific isolation he was beginning to feel in Karlsruhe, and because he saw that Hermann was revitalizing the Observatory. When Otto first settled in Karlsruhe, there had been two other astronomers there, Valentiner and Ristenpart. They were the staff of that Ducal Observatory in Mannheim for the directorship of which Wilhelm had applied some eighty years previously. The building had fallen into such bad repair that the astronomers were temporarily relocated in the capital of Baden-Wurttemberg at Karlsruhe. Valentiner was able to persuade the Grand Duke to build a new observatory in the old university town of Heidelberg and, about this time, they moved to the newly founded Königstuhl Observatory. Otto was asked to make a speech at its formal opening in 1897 and wrote to Newcomb:

> So much I can, perhaps, still perform, although the task is not quite easy, since it is irrelevant to speak of the expectation of great performances or discoveries by this institute, since its instrumental resources cannot much compete with those of other recent observatories, especially those in your fatherland. However, something can be said of an expected yield from such small institutes, if only the right man stands at the helm and knows how to make the available means productive.

> Something of the sort I hope from my son Hermann, who also in Königsberg has only comparatively feeble instrumental resources at his disposal.... While I speak of Hermann, it pleases me to be able to tell you that, as Loewy a little while ago announced to me, he has been recognized by the Paris Academy with this year's Damoiseau Prize.[19]

Photograph of Otto on his eightieth birthday (Schubert, Karlsruhe). (Reproduced from a print held by the Library of Congress, Cleveland Abbe Papers, Container 19).

Otto never gave his speech in Heidelberg since, once again, he went back to Russia. Döllen died about this time (1897), and Otto felt the loss very keenly. For the first time his letters to Newcomb were confused. He wrote as if Döllen were still alive and refers to the deaths of his sisters Charlotte Döllen and Olga Lindhagen as if they were recent, although each of these ladies had died three years earlier. In August 1897, Tisserand, the Director of the Paris Observatory, also died and Otto was called upon to perform one more duty for the *Carte du Ciel*. The *Comité permanent* had to authorize the new Director of the observatory in Paris to act also for them. Otto sent a circular letter to the members of the *Comité*. The copy sent to Gill was accompanied by a personal letter which, in contrast to that written to Newcomb, is beautifully clear and lucid. It is written in English and, except for a few words obliterated by blots, is easy to read. It gives an account of both his public and private concerns in that year:

My dear friend,

Once more I am asked to help to give a legal character to the activity of the Director of the Paris Observatory, as President of the Comité permanent de la Carte photographique du Ciel. With this view I send you the enclosed circular and avail myself of this opportunity to remember me to you and your dear Lady.

During the last two years I have heard from you by way of America through Newcomb and Jacobi [sic, presumably Jacoby is meant]. Also I have learned by you[r] publications that you continue to work with the usual [industry] and also [with su]ccess. --Of course I cannot but regret that last year, after and during the Paris meeting of the Comité permanent, you have been and passed near Karlsruhe without giving me notice of your being so near, or, what I had wished yet more, without visiting us here. I am sorry to hear that your dear Lady's health has lately considerably suffered, and would be happy if you could inform me that she is now completely restored to health.

[Do] not wonder that I do not write more frequently and with more details. Astronomically I am allmost [sic] out of the world and have nothing to communicate.

My health is on the whole very satisfactory and we enjoy very much the better climate in the South of Germany. My sons Hermann (at Königsberg) and Ludwig (at Charkow) work hard in Astronomy and I trust you will hear from time to time of the products of their work. The last named son (Ludwig) has married last year and a couple of days ago his wife has given him a son. Almost the same day I have been informed that a daughter of my second son (Alfred) has also been delivered of a son. Thereby I am now for the first time promoted to the hon[orary rank] of Urgrossvater. I am ashamed to ack[nowled]ge that [I do] not certainly know what is the English expression for that dignity, perhaps Archigrandfather?

Through the Astronomische Nachrichten you will have been informed in the spring of the decease of our dear Döllen. For me personally the loss is immense; during 60 years his life and thinking had formed part of my own being.

Otto, Emma and Eva. A photograph taken in Karlsruhe two days before Emma died in 1902.

Remember me, dear Gill, most kindly to your Lady and let me hear from you, though only by [few] lines Do not forget

Your old friend

Otto Struve

Excuse my bad English writing Since years I am quite out of practice [20]

Thus was the birth of the second Otto Struve announced to the astronomical world. The old man apparently did not know, when he wrote this letter, that Ludwig's son would bear the name of Otto. Certainly, no-one could have guessed that this new baby was to bring greater lustre to the Struve name than any other astronomer in the family since Wilhelm himself.

Newcomb developed the habit of writing to Otto on his birthday: in 1889, for Otto's eightieth birthday, Newcomb telegraphed his greetings. Later that summer, for the last time, the two friends saw each other when Newcomb came to Europe and visited Karlsruhe. From the following year, 1900, Otto's letters became noticeably more repetitive, and even complaining in tone. His sight was failing, so his principal pleasures --reading and writing-- became progressively more difficult. Nevertheless, he planned to take Emma and Eva to Russia at Whitsuntide, 1900, and even spoke of going to the *Astronomische Gesellschaft* meeting in Heidelberg --but eventually decided that he felt himself no longer "fit for that kind of assembly".[21]

Early in March 1902, Emma died quite suddenly and Otto became a widower once more. Amazingly, he took Eva to St Petersburg that summer --for her health, he said-- and enjoyed being surrounded by "children, grandchildren and great-grandchildren".[22] This journey deprived Otto of his last chance to see Newcomb who was, once again, in Europe, but it gave him another opportunity to visit Hermann in Königsberg and to satisfy himself that his son was still reviving the work of Bessel's once-great observatory. By November, Otto was back in Karlsruhe, but the winter of 1902-3 proved trying for him. On account of weakness in the feet, he could take "only a few steps"[23] outdoors, on the best days. His eyesight and hearing continued to deteriorate, and his letters became particularly hard to read.

Despite these problems, Otto and Eva went to Königsberg again in the summer of 1903. Eva was taken ill there, however, and when she recovered, Otto fell ill of a gall-stone. The pair fled back to Karlsruhe to consult their own physician, who recommended a "systematic cure in Baden-Baden".[24] Otto's last surviving letter to Newcomb is undated, but appears to have been written towards the end of 1904. In it, he refers to Hermann's appointment as the Director of the Berlin Observatory --Hermann was charged with moving it to a new site and with completely renewing its equipment-- but, surprisingly, Otto never mentioned to Newcomb the award to Hermann of the 1903 Gold Medal of the Royal Astronomical Society, the Struve family's third such medal. This was awarded, of course, for Hermann's work on the satellites of Saturn. Otto was even able to manage a little humour at his own expense:

My bodily state is in general very satisfactory if [one] abstracts from it that eye and ear only very weakly function and that through weakness of my feet I am condemned to almost complete immobility. Eva is again completely restored from her knee-problem and thanks you very much for your kind remembrance.[25]

Otto died on April 14, 1905, less than a month before his 86th birthday. Newcomb had already mailed his customary greetings, as Hermann acknowledged some months later, giving also some details of his father's death:

A family party in Königsberg in 1902.   Everyone in the picture is descended from Jacob Struve.   Left to right: Elisabeth (Hermann's daughter), Otto, Eva, Georg (Hermann's son), Olga (Hermann's wife and second cousin), Hermann, Berta and Hedwig (sisters of Olga).   From a print provided by Dr. Wilfred Struve.

I saw my father last in the autumn of last year, before I moved from Königsberg to Berlin. I found him at that time already very frail and dependent on his companions for everything. What grieved him most was the weakening of his eyesight, which more and more hindered him in his usual occupation at the writing desk. He was, however, even in the course of last winter, spared pain and suffering, and only two days before his death, the first heart-disturbance showed itself in a slight breathing difficulty, which by its repetition prepared him for a gentle end.

My sister Eva is staying first for some months in St Petersburg with my eldest brother; later she will probably come to us in Berlin and there look around for some occupation.[26]

One cannot but admire Eva who, as youngest daughters often did in those days, gave up her own life to care for her aging parents. Otto became rather a pathetic figure in his last years, and death did not come to him too soon. Eva must have found him difficult to live with, particularly after her mother died. When Otto himself died, Eva was 32 and had nearly another half-century to live. Later she was joined by her cousin, Hermann's sister-in-law Hedwig, and these two women survived both world wars in Berlin, holding together the various branches of the family as long as they both lived.

Memorial marking the graves of Otto, his wife Emma, daughter Eva and cousin Hedwig Struve.

Otto's interests in astronomy had been very wide, but double-star measurement and exact positional astronomy remained his principal loves. In the former, he was probably his father's equal and modern astronomers would be grateful to him if he had done nothing else. Nevertheless, the spark of inspiration so evident in Wilhelm's work seems to be missing from Otto's and, despite his solid wide-ranging contributions, the son does not have quite the same standing in the history of astronomy as his father had. This was partly because of the toll of administration, and not the least of Otto's contributions to astronomy was relieving his father of so many chores so that *he* could concentrate on research. In later life, Otto received the deference due to a senior observatory director who was in charge of one of the world's major observatories. Perhaps his nearest twentieth-century equivalent is the kind of man who can obtain many research grants and thus attract younger workers whose work further increases his own prestige. He knew how to find his way through the Imperial civil service and not only kept Pulkovo afloat, but also founded a new observatory in Tashkent. His letters reveal more of his character than do his father's of Wilhelm's. Otto seems the harsher and hastier in his judgments --he never appears, for example, to have been reconciled with Backlund-- but his sense of humour more often shines through than does Wilhelm's. One can still enjoy many chuckles reading Otto's letters; even some of his remarks critical of other astronomers may have originally been private jokes. Otto was loyal to friends, and these included Airy, Le Verrier, Newcomb and Gill --all men whose names have lived with his in the history of astronomy.

By contrast, those whom he dismissed, even if somewhat unfairly, have been less well remembered. If his judgments have been vindicated by history, he must have been a man of insight.

According to Nyrén, Otto's wish was that his ashes should be buried under the pier of *his* instrument, the 15-inch refractor at Pulkovo.[27]   When Nyrén wrote, the intention was to carry out this wish, but it was not fulfilled. Otto's body was returned to the Struve ancestral home of Horst and buried there in the churchyard as, in due course, was Eva's. Despite the ravages of time and war, a monument still stands.

NOTES

1.   O.W. Struve, 1885, letter of February 26 to Simon Newcomb.
2.   M. Nyrén, 1905, <u>Vierteljahrsschrift der Astronomische Gesellschaft</u>, Vol. 40, pp. 286-303 (English translation, 1906, in <u>Popular Astronomy</u>, Vol. 14, pp. 352-368).
3.   K.H. Struve, 1890, letter of 15/27 November to Simon Newcomb.
4.   O.W. Struve, 1889, letter of March 30 to Simon Newcomb.
5.   O.W. Struve, 1889, see Chap. 14, ref. 2.
6.   D. Gill, 1889, letter of July 16 to O.W. Struve (RGO 15/126 ff. 846^r - 848^r).
7.   O.W. Struve, 1889, letter of September 2 to D. Gill.
8.   O.W. Struve, 1889, letter of May 24 to Simon Newcomb.
9.   M. Nyrén, 1905, see ref. 2.
10.   O.W. Struve, 1890, letter of April 28-May 1 to Simon Newcomb.
11.   O.W. Struve, 1891, letter of October 3 to Simon Newcomb.
12.   O.W. Struve, 1893, letter of March 17 to Simon Newcomb.
13.   D. Gill, 1893, letter of August 9 to O.W. Struve (RGO 15/127 ff. 337^r - 338^r).
14.   O.W. Struve, 1895, letter of April 16 to Simon Newcomb. Translations of parts of this letter and others quoted here have been published by K. Krisciunas, 1984, <u>Quarterly Journal of the Royal Astronomical Society</u>, Vol. 25, pp. 301-305, who also discusses the events surrounding Otto's resignation.
15.   O.W. Struve, 1898, letter of April 26 to G.V. Schiaparelli (see ref. 16)
16.   G.V. Schiaparelli, 1898, letter of May 1 to O.W. Struve. (This letter and the preceeding one have been published in <u>Corrispondenza su Marte di Giovanni Virginio Schiaparelli</u>, Vol. II (1890-1900), 1976, Domus Galileana, Pisa, pp. 227-8, and are quoted with permission.
17.   O.W. Struve, 1896, letter of May 13 to Simon Newcomb.
18.   O.W. Struve, 1896, letter of November 19 to Simon Newcomb.
19.   O.W. Struve, 1897, letter of November 19 to Simon Newcomb.
20.   O.W. Struve, 1897, letter of August 14 to D. Gill (RGO 15/129 ff. 376^{r-v}).
21.   O.W. Struve, 1900, letter of April 10 to Simon Newcomb.
22.   O.W. Struve, 1902, letter of July 6 to Simon Newcomb.
23.   O.W. Struve, 1902, letter of November 23 to Simon Newcomb.
24.   O.W. Struve, 1903, letter of October 18 to Simon Newcomb.
25.   O.W. Struve, 1904?, last undated latter to Simon Newcomb in answer to one from Newcomb dated 1903 November 2, which Otto had "left unanswered for a long time".
26.   K.H. Struve, 1905, letter of August 20 to Simon Newcomb.
27.   M. Nyrén, 1905, see ref. 2.

EPILOGUE

THE FAMILY TRADITION

By the time that Otto died, his two astronomer sons, Hermann and Ludwig, had completed the most active parts of their own careers. These two, the third generation of astronomers in their family, were the fourth generation in the scholarly tradition, of which their great-grandfather Jacob can justly be regarded as the founder. His ancestors, brothers and sisters were, no doubt, all worthy people, but they would be forgotten if Jacob's skill at the organ and talent for arithmetic had not singled him out for an education. Many of his descendants, however, in the fifth, sixth or even seventh generation, have been men and women of intelligence and achievement. Such family traditions are not without parallel: the Herschels, whose story is intertwined with that of the Struves also formed a dynasty whose founder was undoubtedly an even more eminent astronomer than Wilhelm Struve --although the Struves perhaps conserved their excellence through more generations. Other lesser dynasties, Fuss and Knorre, have been noted in the preceding chapters, and examples could be mustered from other sciences (e.g. the Becquerels in physics[1]) and from the arts. Returning to astronomy, the Cassini family --of Italian origin-- dominated French astronomy as members of four successive generations became Directors of the Paris Observatory.

Hermann and Ludwig each survived their father by only fifteen years. After moving to Kharkov, Ludwig became involved in university administration --particularly the founding of a school of engineering-- and made no more major contributions to research. He did teach his son, the second Otto, and Boris Gerasimovich, a later Director of Pulkovo who disappeared in Stalin's purges. Hermann's job in Berlin was to build and to equip a new observatory --also a Struve tradition-- away from the city centre. In this task he fully justified his father's hopes for him, but inevitably was able to do less research himself. Earlier plans to move the observatory had foundered on lack of funds and conflicts of personality. Hermann could rise above the latter in Berlin in a way that would have been impossible for him in Pulkovo, and sold the old site so advantageously that, like his grandfather, he could build and equip the new observatory (at Berlin-Babelsberg) in a style "worthy of an Emperor", but not "displeasing to a Minister of Finance" (Chapter 7). The last instruments were removed to their new home in 1913, but the buildings were not completed until after the outbreak of the First World War. Hermann continued the family correspondence with Newcomb at least until 1908 (Newcomb died the following year) in which year an international congress of mathematicians brought the American to Rome. Hermann also tried to arrange for Eva (also in Rome that year) to meet Newcomb and wrote (in English):

Hermann in 1904. (Reproduced by permission of the Royal Astronomical Society.)

Ludwig: photograph of unknown date and provenance.

I can not yet decide, if it will be possible for me to go abroad this summer, because I am occupied with the affairs of our observatory, the plan of transferring the old observatory to a better place, going forward somewhat slowly by different causes But it may be, that we might bring our daughter Lita, now 15 years old, who was during the last winter very ill with a long nervous sickness, for recreation to Switzerland, and than [sic] of course I will try to visit you there [i e Rome] [2]

The First World War was a difficult time for the family, now split between warring nations. Before the war was finished, the Russian revolution increased Hermann's concern for his brothers and sisters still in Russia, a concern which is believed to have hastened his own death In 1919 his wife Olga died and, about a year later, he broke his thigh in a fall from a tramcar While he was apparently recovering in Herrenalb, a spa in the Black Forest, a long-standing weakness of the heart led to his death on 12 August, 1920.[3]

Hermann's concern for the Russian part of the family had been fully justified His father's cousin Peter --son of the "resolute and undertaking character" Bernhard-- was politically very active. Perhaps the most outstanding of the Struves (apart from the astronomers), Peter knew Lenin and introduced him to Marxism. Later, these two men quarrelled and Peter Struve moved gradually across the political spectrum ending up as a political adviser (especially in foreign affairs) to the White Russian Generals Denniken and

The second Otto Struve during a visit to Victoria (Dominion Astrophysical Observatory photograph by S H Draper)

Wrangel (the latter of a family long associated with the Struves). After the collapse of the White Russian army, Peter left Russia and ended his days in Paris (after internment by the Germans in the Second World War) a devout son of the Russian Orthodox Church.[4] He always hoped that one of his five sons would become an astronomer, but none did.[5]

Ludwig also suffered during the Revolution. His eldest son, the second Otto, had not been of military age at the outbreak of the First World War but, nevertheless, travelled to St Petersburg in 1915 to volunteer. He would not have been accepted but for his great height (a family trait) which qualified him for a guards' regiment and attracted the attention of a recruiting officer. He received his commission in the Imperial Army in February 1917 and was among the last officers to be promoted by Imperial decree.[6] Otto served actively on the Turkish front until the collapse of the Provisional Government in 1917, when he returned to Kharkov and graduated in 1919. He then joined the White Russian army. Ludwig and his family suffered considerable privation during the civil war, and were obliged to leave Kharkov for the Crimea. There Ludwig was appointed as a professor in the Tauride University in Simferopol, but the difficulties leading up to his flight from Kharkov had taken their toll. He died in the new university during the course of a lecture, in November 1920.[7] He may never have heard of the death of his brother Hermann, a few months earlier.

Another book would be needed to do justice to the second Otto's career. With the rest of the White Russian army, he was forced to retreat into Turkey and was interned for a while in Gallipoli. After being released he went to Constantinople (Istanbul) where he took any job he could find. The then Director of the Yerkes Observatory of the University of Chicago, E.B. Frost, had heard that Ludwig's son had graduated as an astronomer and learned the younger Otto's whereabouts from Eva von Struve.[8] She probably had the information from her cousin Peter Struve, who had travelled through Germany from one crumbling White Russian front to another. Peter's son, Gleb Struve, who lived in California until his recent death, has written of his father's memories of Otto working as a boot-black in the streets of Constantinople.[9] Frost wrote to Otto, and despite the uncertainties of the times, the letter reached the intended recipient --although Otto had to go to the Y.M.C.A. in Constantinople to find someone to translate it. The letter offered him a job, and in October 1921 Otto went to Chicago to continue his studies under Professor Frost. Later, Otto succeeded Frost as Director of the Yerkes Observatory of the University of Chicago and founded both the McDonald Observatory of the University of Texas, and the National Radio Observatory in Virginia --thus following a family tradition of founding observatories. He departed from the family tradition of exact positional astronomy, however, and made major contributions to astrophysics. He was elected to the U.S. National Academy of Sciences in 1937, and received the Struve family's fourth Gold Medal from the Royal Astronomical Society in 1944 for his work on astronomical spectroscopy. No other family has received this honour in four consecutive generations. From 1932 to 1947, Otto was editor of the *Astrophysical Journal*. In 1952, Otto was elected to a three-year term as President of the International Astronomical Union --surely partly a recognition of the efforts in international cooperation made by his grandfather and his

Uncle Hermann (who had participated in the fourth conference of the International Union for Cooperation in Solar Research, held in the U.S.A. in 1910). In the late 1950s, illness took its toll of the second Otto and he died in April 1963.

Although he married, Otto died childless. His brother (who also studied astronomy) and sister both died during the Russian civil war, so Ludwig's branch of the family is extinct. Other members of the family who made minor contributions to astronomy include Karl (or Kyrill the ambassador) and a daughter of Hermann's brother, Alfred, who was a computing assistant at Pulkovo Observatory. Eva Struve also performed similar work at the Babelsberg Observatory.

The most important astronomer in the family still to be discussed, however, is Hermann's son Georg. More than two years older than his better known cousin, Georg was already a qualified astronomer at the outbreak of the First World War. He had studied at the Universities of Heidelberg and Berlin and, after six months compulsory service with the marines, received an appointment as a civilian observer at the Marine Observatory in Wilhelmshaven --where he remained throughout the 1914-18 war. In 1919 he moved to the new Babelsberg Observatory and remained there for the rest of his life. He continued in the Struve tradition of positional astronomy and particularly in his father's study of the satellite systems of Saturn and Uranus. By this time, Saturn was in the southern sky and, to observe it more easily, Georg travelled first to the Lick Observatory in 1925 (meeting his cousin Otto while he was in the U.S.) and to the former Union Observatory in Johannesburg in 1926. The late W.S. Finsen wrote of Georg's amazement at the steadiness of the telescopic images seen in Johannesburg.[10] Contemporaries praised Georg's skill as an observer, and it was scarcely possible to improve on the knowledge of Saturn's satellites, obtained by Georg and Hermann together, until space probes went right through the system. From Johannesburg, Georg returned to Berlin to prepare his work for publication. These were the years leading up to Hitler's rise to power and Georg, although not a member of the Nazi party, was politically very active. This activity, combined with the hard work in his profession, proved too much for him and he was taken ill early in 1933. Although at first expected to recover, he died suddenly before his forty-seventh birthday.[11]

Georg had two sons, Wilfried and Reinhardt. The elder was studying astronomy at Heidelberg at the time of his father's death, completed his doctorate and was, for a while, a volunteer assistant at the Babelsberg Observatory founded by his grandfather. Both brothers were conscripted into the German army at the outbreak of the Second World War, and --ironically-- each found himself on the Russian front. The younger brother did not survive the war, and the elder gave up the study of astronomy. The principal line of descent from the Struve astronomers continues through him, however, since he now has grandchildren bearing the Struve name.[12]

Two of Wilhelm's daughters married astronomers. The family *Ahnenlisten*[13] records that Döllen, Charlotte's husband, had a great-grandson, Erich Lange, who became an astronomer. A man of this name worked at the

Georg Struve at the former Union Observatory, Johannesburg. Georg is first from left on the front row. On his right is the famous double-star observer R.T.A. Innes. At right, on the back row, is W.S. Finsen who provided the print from which this is copied.

German Hydrographic Institute in Hamburg until his death in 1961. D.G. Lindhagen, Olga's husband, had three sons, one of whom --Arvid-- was a well-known Swedish astronomer interested primarily in calendrical questions. Many other members of the family achieved distinction in other fields --mostly scholarly ones. To try to trace them all would produce a catalogue unlikely to interest the general reader. An account of many branches of the family was included by Wilhelm Henop in his 1931 article *Der Gymnasialdirektor Jacob Struve und die Seinen.*[14] Henop, a medical doctor in Altona, was a grandson of Wilhelm's older sister Christiane and a nephew of Kasimir Henop, employed by the Repsolds, who prepared drawings for some of the Pulkovo instruments. The genealogy of the family was brought up to date by Olga Tomaschek, great-granddaughter of Wilhelm and granddaughter of Anna Cremers, who compiled the *Ahnenlisten der Astronomenfamilie Struve*[13] as completely as she could up to 1960. Both documents are rare but extant, and the interested reader should look for them.

The persistence of so strong a tradition of astronomical research through five generations of a family invites speculation. In the nineteenth century it was, of course, more common for a son to follow his father's calling than it is now and, earlier still, Imperial Russia had laws requiring men in certain social classes to engage in their father's occupations.[15] Such traditions were natural in an hereditary aristocracy, but it is hard to make one's son into an astronomer if he does not wish to be one, and impossible to make him into a good astronomer unless he has the ability; yet at least four of the Struves were outstanding astronomers.

We do not know if Wilhelm deliberately cultivated an interest in astronomy in his oldest surviving son, or whether Otto's brother Alfred, if he had lived, would have been groomed to succeed his father. Wilhelm appears to have been content to let his other sons choose their careers, which suggests that he probably did not try to force Otto into a mould. Nevertheless, the journeys to Europe to order instruments for Pulkovo (Chapter 6) must have influenced Otto. He himself dates his interest in astronomy to the founding of Pulkovo.[16]

Similarly, Otto did not appear to force his sons into astronomy. He took considerable pride in the achievements of Georg Wilhelm and Alfred (his son, *not* his brother) and nowhere expressed regret that they did not become astronomers. He also seemed reconciled to Hermann being a mathematical physicist. Once more, it was a new instrument in Pulkovo that excited a young Struve and brought him into astronomy. Perhaps Otto had that result in mind when he took Hermann to America, but there is no sign that he put pressure on his son to become an astronomer. Ludwig's decision also appears to have been spontaneous.

In the fourth generation, however, there is some evidence that the tradition was becoming self-conscious, for example in unpublished remarks by the younger Otto.[17] A picturesque symbol of this may be the *Beobachtungskäppchen*.[18] Characteristic in portraits of both Wilhelm and Otto are the little "pill-box" hats they wear. The family tradition is that the bride of a Struve astronomer should make such a cap for her new husband. The males of the family tend to go bald early and the caps, made out of red velvet were to protect their heads when they were observing. Presumably, Emilie Wall did this spontaneously for Wilhelm, and round the cap she sewed one gold thread. Otto had a similar cap made, it can be assumed, by his Emilie; it had three gold threads. By the time Hermann married, a rule had been codified; the number of gold threads was to indicate to which generation of astronomers a Struve belonged. So Hermann also had three threads. I do not know if the younger Otto had such a cap, but Dr. Louis Berman, who was Georg's observing assistant at the Lick Observatory recalls that Georg did have a cap with four gold threads.[19] A family photograph, taken in Königsberg in 1902, shows both Hermann and the older Otto wearing their *Beobachtungskäppchen*.

Modern times have seen three great invasions of European Russia by her western neighbours, each largely unprovoked, and each of considerable significance to the Struve family. If Wilhelm had not escaped his French captors, he might well have marched into Russia as a private soldier, and we should never have heard of him. Yet, paradoxically, it was Napoleon, whom the Struve family so heartily disliked, who created the conditions that led to Wilhelm's migration to the Russian empire and the foundation of his family there. When the second invasion was led by the Kaiser Wilhelm and the Emperor Franz-Joseph, the Struves' greatest contributions to Russia had been made, and part of the family had returned to Germany and was becoming German in its outlook. Two astronomer cousins, Georg and the younger Otto, were on opposing sides, although the former had no combatant duties. The last invasion, Hitler's, symbolized the end of the Struve dynasty as a major force in astronomy, even though the younger Otto lived and worked for nearly twenty years after the end of the Second World War. One of the grimmest battles of that war was the year-long siege of Leningrad (the old St Petersburg) and Pulkovo was marked out for destruction on account of its commanding view of the city. The Observatory was completely razed, and although you can still visit it today and even go into the very rooms that Wilhelm and Otto described, they are reconstructions on the old foundations. Pulkovo was rebuilt soon after the end of the war as a symbol of national reconstruction in the Soviet Union. But the real memorial to this family is the outstanding contributions made to the science of astronomy by those resolute and undertaking characters, Wilhelm and Otto Struve.

Postage stamps commemorating Wilhelm Struve, issued by the Soviet Union. Left: A 40-kopek stamp issued in 1954 to mark the reconstruction of Pulkovo Observatory. It shows Wilhelm ("Founder of Observatories") flanked by Bredichin ("Distinguished astronomer, investigator of comets") and Belopolsky ("Pioneer of Russian astrophysics"). Right: A 4-kopek stamp issued in 1964, on the 100th anniversary of Wilhelm's death. Wilhelm's name is Russianized as Vassily Yakovlevich, hence the initials that, transliterated, read V.Ya.

NOTES

1   Abraham Pais, 1986, Inward Bound, Oxford University Press, pp 44-5

2   K H Struve, 1908, letter of April 8 to Simon Newcomb

3   L Courvoisier, 1920, Astronomische Nachrichten, Vol 212, pp 33-7

4   R E Pipes, 1970 and 1980, Struve, Liberal on the Left 1870-1905 and Struve, Liberal on the Right,1905-1944, Harvard University Press

5   Gleb Struve, 1980, private communication, March 19

6   Otto Struve, 1941, unpublished autobiographical notes deposited with the Archives of the U S National Academy of Sciences, cited with permission

7   L Courvoisier, 1921, Astronomische Nachrichten, Vol 212, pp 351-2

8   E B Frost, 1921, Popular Astronomy, Vol 29, pp 536-541, see also the same author's An Astronomer's Life, Houghton Mifflin and Co , Boston, N Y , 1933, pp 255-6

9   Gleb Struve, 1980, see ref 5

10   W S Finsen, 1975, private communication, July 11

11   J Dick, 1934, Vierteljahrsschrift der Astronomische Gesellschaft, Vol 69, pp 2-8

12   W Struve, 1977, private communication, March 2

13   O Tomaschek, 1960, Ahnenlisten, p 7

14   W Henop, 1931, Jacob Struve, pp 15-18

15   R E Pipes, 1974, Russia under the old Regime, Weidenfield and Nicolson reprinted 1982 by Penguin Books, Hammondsworth and New York, pp 87, 100

16   I am indebted to W R Dick of Potsdam Observatory for telling me of the existence at Kharkov Observatory of unpublished autobiographical notes of the elder Otto and for supplying me with this information from them

17   Otto Struve, 1941, see ref 6

18   W Struve, 1977, in a letter of May 19 gives us the story of the Beobachtungskäppchen, parts of which are also given in ref 10 and 19

19   L Berman, 1977, private communication, December 17   Georg's cap was also mentioned by Finsen (ref 10)

NAME INDEX

German names with "von" are listed alphabetically under the main surname, e g , Alexander von Humboldt is found under "Humboldt, Alexander v " Not everyone entitled to use "von" did so consistently, so it may not always appear in the Index In particular, "von" has been omitted from all Struve names, even though Wilhelm and all his descendants were entitled to use it Similarly, French names that take "de" are listed under the surname, but note that the English astronomer "de la Rue" is found under "D"

Married women are listed under their married surnames, but cross-references are given under their maiden names whenever these have also been used in the text

Members of reigning houses are given under their usual names, e g , the Kings of Prussia are found under "Friedrich Wilhelm"

Relationships given in parentheses with the dates are to Wilhelm Struve, unless otherwise specified

------

Abbe, Cleveland (1836-1916) American astronomer and meteorologist who visited Pulkovo 36, 68, 72, 105-7, 164, 169 passim, 196, 199-201, 221
Adams, John Couch (1819-1892) British astronomer who successfully predicted the position in which Neptune would be found 135-6
Airy, Annot (fl 1865) daughter of George Biddell Airy 163, 194
Airy, Christabel (fl 1870) daughter of George Biddell Airy 193-4
Airy, George Biddell (1801-1892) Seventh Astronomer Royal of England from 1835-1881 59, 85 90-101, 102, 109, 117, 131 passim, 146, 150-54, 156 passim, 163 passim, 186 passim, 209, 211, 217, 233
Airy, Richarda (née Smith 1804-1875) wife of George Biddell Airy 91, 92, 96-7, 163, 193
Airy, Wilfrid (fl 1890) son of George Biddell Airy (who also had a nephew of the same name) 163, 194
Aitken, Robert Grant (1864-1951) well-known American observer of double stars, Director of the Lick Observatory 139
Albert, Prince Consort of Queen Victoria (1819-1861) 136
Albert Edward, Prince of Wales (1841-1910) King Edward VII of Great Britain from 1901 195
Alexander I (1777-1825) Tsar of Russia from 1801 13, 17, 36, 49, 67, 219
Alexander II (1818-1881) Tsar of Russia from 1855 67, 101, 107, 157, 160, 164, 171, 187
Alexander III (1845-1894) Tsar of Russia from 1881 213, 220
Alexandra Feodorovna (fl 1830) Tsarina to Nicholas I 78
Ångström, Anders Jonas (1814-1874) Swedish physicist who gave his name to the unit used for measuring wavelengths of light 174
Anna (1693-1740) Empress regnant of Russia from 1730 36
Anne, St (fl 14th century) Russian Duchess, later canonized and patron saint of Russian order of chivalry 97

SUBJECT INDEX